AMTE Monograph Series Volume 7

Mathematics Teaching: Putting Research into Practice at All Levels

Edited by

Johnny W. Lott
University of Mississippi

Jennifer Luebeck
Montana State University

Monograph Series Editor

Marilyn E. Strutchens
Auburn University

Published by

Association of Mathematics Teacher Educators
San Diego State University
c/o Center for Research in Mathematics and Science Education
6475 Alvarado Road, Suite 206
San Diego, CA 92129

www.amte.net

Library of Congress Cataloging-in-Publication Data

Mathematics teaching : putting research into practice at all levels / edited by Johnny W. Lott, Jennifer Luebeck.
p. cm. -- (AMTE monograph series ; v. 7)

ISBN 978-1-62396-953-0

1. Mathematics teachers--Training of--United States. 2. Mathematics--Vocational guidance--United States. I. Lott, Johnny W., 1944– II. Luebeck, Jennifer. III. Association of Mathematics Teacher Educators.

Mathematics teaching : putting research into practice at all levels / edited by Johnny W. Lott, Jennifer Luebeck.

QA10.5.M42 2010
510.23'73--dc22

2010037371

The publications of the Association of Mathematics Teacher Educators present a variety of viewpoints. The views expressed or implied in this publication, unless otherwise noted, should not be interpreted as official positions of the Association.

Contents

Reys, B. J.
AMTE Monograph 7
Mathematics Teaching: Putting Research into Practice at All Levels

Foreword

On behalf of the Association of Mathematics Teacher Educators (AMTE), I am pleased to introduce this important resource. The seventh monograph of AMTE, *Mathematics Teaching: Putting Research into Practice at All Levels,* edited by Johnny W. Lott (University of Mississippi) and Jennifer Luebeck (Montana State University), highlights examples of important scholarship of and for the mathematics teacher education community.

This monograph, like others produced by AMTE, serves as a forum for mathematics teacher educators to exchange ideas, experiences, resources and detailed accounts of work to improve teacher preparation. Chapters in this monograph take up a variety of issues such as using online social networking in the preparation of teachers, examining the impact of textbook-specific professional development, and offering a mathematics-specific reading in the content area course.

AMTE is pleased to support the dissemination of knowledge important to the field. While the monograph series has served as an important vehicle, we realize that what is needed is a more frequent and accessible outlet for the knowledge accumulated by the field. AMTE is responding to this need by initiating a practitioner-based scholarly journal focused on mathematics teacher education. The journal will contribute to building a professional knowledge base in mathematics teacher education that stems from practitioner knowledge. The new journal will be jointly published by AMTE and the National Council of Teachers of Mathematics (NCTM) and the first issue is scheduled for release in 2012.

AMTE is also partnering with the editors of the *Journal of Mathematics Teacher Education* (JMTE) to publish a special issue of the journal focusing on equity issues in the mathematics education of teachers. The special issue is edited by Marilyn Strutchens and due out this year. This issue is especially important given the growing numbers of diverse learners in mathematics classrooms, and the need to understand how to prepare mathematics teachers that can effectively eradicate the achievement gap and diminish other related disparities in mathematics education.

As noted, AMTE dissemination efforts are expanding, due in large part to the consistently high quality of the AMTE monograph series. This present installment is no exception and further solidifies both the need for and the quality of current work by the mathematics teacher education community. With the launch of the new AMTE journal, the monograph series will become an "occasional" rather than annual publication. That is, as particular issues warrant, monographs will be commissioned to address and/or report to the community. In this way, AMTE can continue to disseminate important information and findings through multiple venues.

On behalf of AMTE, I thank those involved in the development of this seventh AMTE monograph including:

Co-Editors:

Johnny W. Lott, *University of Mississippi*
Jennifer Luebeck, *Montana State University*

Editorial Panel:

Jane Keiser, *Miami University of Ohio*
Carol Malloy, *University of North Carolina*
Eric Milou, *Rowan University*
Melfried Olson, *University of Hawaii*

Laura Spielman, *Radford University*
Sheri Stockero, *Michigan Technological University*
Amy Hillen, *Kennesaw State University*
Dorothy White, *University of Georgia*
Trena Wilkerson, *Baylor University*

AMTE Monograph Series Editor
Marilyn E. Strutchens, Auburn University

As a final note, on behalf of the AMTE Board I'd like to extend a special thanks to Marilyn Strutchens, Monograph Series Editor, for the two most recent monographs as well as the special issue of JMTE. With these publications, Marilyn is completing her responsibilities as Series Editor — just in time to prepare to take on another leadership role in January 2011 as President of AMTE. Marilyn's steady leadership with regard to these publications has contributed to the quality and timeliness of the work. It is these qualities that will also serve the organization well as she assumes her new role.

Barbara J. Reys
AMTE President 2009–2011

Lott, J. W. and Luebeck, J.
AMTE Monograph 7
Mathematics Teaching: Putting Research into Practice at All Levels

1

Mathematics Teaching: Putting Research into Practice at All Levels

Johnny W. Lott
University of Mississippi

Jennifer Luebeck
Montana State University

In *Scholarship Reconsidered: Priorities of the Professoriate*, Boyer (1990) identified four kinds of scholarship corresponding to discovery, integration, application, and teaching. Though Boyer's work caused much controversy by calling for the university professoriate to rethink how research, teaching and service are considered, it does provide a framework for thinking about putting research into practice at all levels, including in the mathematics classroom. As a creative, if imperfect, means of organizing this monograph, we have attempted to draw parallels between Boyer's four categories and the scholarship presented by our fourteen authors.

Boyer's first category, the *scholarship of discovery,* characterizes the type of small-scale research often conducted in the classroom setting. The discovery scholarship recounted in this section seeks to directly influence teaching and learning through experimenting with classroom practice—in this context, the practice of teacher educators. Kosko, Norton, Conn and San Pedro describe an experiment designed to enhance a mathematics content course. Lenges, and van den Kieboom and Magiera explore instructional stances that may enhance preservice teachers' mathematical understanding. Arbaugh, Lannin, Jones, and Barker extend discovery scholarship to the context of professional development.

In Boyer's terms, the *scholarship of integration* deals with "making connections across the disciplines, placing the specialties in larger context, illuminating data in a revealing way" (1990, p. 18). Mathematics and education meld in the process of integrating scholarship, allowing researchers to interpret their findings in a broader intellectual context. In this volume, integration is represented by the blending of teaching and research as discussed by van Zoest, Stockero and Edson; the interactions between authority and practice outlined by Mewborn; and the strategies to embed reading in a mathematics course provided by Thompson.

The *scholarship of application* supports the use of knowledge and research to seek solutions for significant issues, including the need for more, and more adequately prepared, mathematics teachers. Benken and Gomez-Zwiep investigate the content component of an alternative certification program in mathematics. Lee, Ives, Starling, and Hollebrands explore the implementation of curricular advances in teaching statistics with technology. Miriti and Mohr-Schroeder demonstrate how technology can be used to enhance the supervision and mentoring of prospective mathematics teachers.

The remaining articles in this monograph—and perhaps all of them—fit best in the category termed the *scholarship of teaching*. In Boyer's words (1990), teaching "not only means transmitting knowledge, but transforming and extending it as well"; it promotes "active, not passive, learning and encourages students to be critical, creative thinkers, with the capacity to go on learning" (p. 23). Finally, "pedagogical procedures must be carefully planned, continuously examined, and relate directly to the subject taught" (p. 24). Chapters by Santagata and van Es, and Suh and Parker describe how they introduce preservice teachers to structured, disciplined, and content-focused analysis of instruction. Leatham and Peterson document their deliberate efforts to redesign the student teaching experience to be more reflective, while Cwikla investigates how classroom video can be used to transform the teaching of college faculty.

It is also possible to draw parallels and distinctions between Boyer's composite view of scholarship and more traditional research practices. As we have learned—sometimes painfully—in mathematics education, research that is not the result of randomized controlled trials has often been viewed as inadequate in federal guidelines, or discounted in reports published by federal offices and organizations (see *randomized clinical trials* at http://ies.ed.gov/ncee/wwc/help/glossary/#gr). The American Statistical Association (ASA) takes a broader view of the efforts of the mathematics education research community, even while urging researchers toward a more cohesive and connected body of work:

> If research in mathematics education is to provide an effective influence on practice, it must become more cumulative in nature. New research needs to build on existing research to produce a more coherent body of work. Researchers in mathematics education are, of course, and should continue to be, free to pursue the problems and questions that interest them. In order for such work to influence practice, however, it must be situated within a larger corpus. School mathematics is an excellent venue for small-scale studies because mathematics learning has many facets, and the classroom is a manageable unit that can be studied in depth and detail. Such studies can cumulate, however, only if they are connected. Studies cannot be linked together well unless researchers are consistent in their use of interventions; observation and measurement tools; and techniques of data collection, data analysis, and reporting. (2007, pp. 4–5)

Clearly, not all of the research referenced in this quote is of the formal experimental variety. Many small-scale studies using the classroom as a manageable unit do not lend themselves to randomization or a control-treatment approach. However, as noted by the ASA, mathematics education will be advanced if, as researchers, we focus on consistent implementations and well-documented, replicable measurement practices that serve to merge, rather than disperse, our knowledge about mathematics teaching and learning.

As a collection of fourteen articles in the Association of Mathematics Teacher Educators monograph series, this volume seeks to promote the implementation of research into practice in classrooms of the mathematics education field. Few of the studies reported in this monograph involve randomized, controlled research designs. They tend to take advantage of existing classrooms, programs, or bodies of students as convenience samples; and as a result, they are more real. While these studies may not report broadly generalizable results, most were developed in a research context and supported by research literature in ways that allow for future replication and extension. They are presented with an understanding that they provide models to be used by others in the field. It is our hope that by building on the contents of this monograph, with similar data collection methods and empirical approaches to teaching, other researchers will add to the cumulative knowledge that is becoming the foundation for excellent mathematics education.

References

American Statistical Association. (2007). Using statistics effectively in mathematics education research: A report from a series of workshops organized by the American Statistical Association with funding from the National Science Foundation. Washington, DC: Author.

Boyer, E. L. (1990). Scholarship reconsidered: Priorities of the professoriate. Washington, DC: The Carnegie Foundation for the Advancement of Teaching.

U. S. Department of Education, Institute of Education Sciences. Glossary of terms. Retrieved from http://ies.ed.gov/ncee/wwc/help/glossary/#gr

---

Johnny W. Lott is professor emeritus of mathematics in the Department of Mathematical Sciences at The University of Montana. He is a retired professor of mathematics and education and former Director of the Center for Excellence in Teaching and Learning at The University of Mississippi. Dr. Lott continues to be involved in curriculum development and teacher preparation.

Jennifer Luebeck is associate professor of mathematics education in the Department of Mathematical Sciences at Montana State University. She works extensively with preservice and inservice mathematics teachers through professional development and undergraduate/graduate programs.

Lee, H. S., Ives, S. E., Starling, T. T., and Hollebrands, K. F.
AMTE Monograph 7
Mathematics Teaching: Putting Research into Practice at All Levels
© 2010, pp. 7–23

2

Knowledge for Teaching Statistics with Technology: Examining Mathematics Teacher Educators' Planning

Hollylynne Stohl Lee
North Carolina State University

Sarah E. Ives
Texas A & M University–Corpus Christi

Tina T. Starling
Karen F. Hollebrands
North Carolina State University

Technology is more frequently being used in the teaching and learning of mathematics and statistics, many times requiring teacher educators to implement new curriculum that incorporates technology. This leads to a need for research on the specialized knowledge needed for effective use of technology. The Technological Pedagogical and Statistical Knowledge framework developed by Lee and Hollebrands was used to study planning and debriefing sessions of mathematics teacher educators' implementation of new curriculum for secondary mathematics teachers. Instances of technological, pedagogical, and statistical knowledge were found when teacher educators extended the curriculum to improve students' understanding as well as to create learning opportunities for students.

Whenever a new curriculum is adopted, teachers must make decisions about (a) how to organize class activities for whole class or small group work, (b) which tasks to pose and key questions to ask, and (c) how text and supplemental resources will be used. Whether resources such as technology will enhance or hinder students' learning depends on teachers' decisions about how and when to use technology tools as aids in teaching and learning mathematics (Lee & Hollebrands, 2008a). Some factors that influence these choices are a teacher's experience with different pedagogical strategies, knowledge of content being taught, and familiarity with different representations and tools.

Through two NSF-funded projects (DUE 04-42319 and DUE 08-17253), teacher education materials have been created to address the need to prepare teachers to use technology to teach mathematics and statistics in ways aligned with recommendations from organizations such as the Association of Mathematics Teacher Educators (2006), the National Council of Teachers of Mathematics (NCTM, 2000), and the American Statistical Association (Franklin et al., 2005). The curricular materials are written as modules with the first module focusing on teaching and learning statistics with technology for middle and high school topics (Lee, Hollebrands, & Wilson, 2010). Though probability and statistics is a common curricular strand in K–12 schools and there has been a rapid increase in students taking the Advanced Placement Statistics exam (College Board, 2007; NCTM, 2000), many teachers only take one college-level statistics course with little attention to pedagogical issues of teaching statistics. The NSF-funded materials can assist mathematics teacher educators in preparing prospective teachers with a deeper understanding of statistical concepts and the use of simulation and data analysis tools.

If new teacher education materials are going to be used effectively with prospective teachers, it is important to consider factors that may affect use of the curriculum. In addition, because many teacher educators are not expected to be experts in the teaching and learning of every topic in mathematics, it is reasonable to believe that some teacher educators may not be comfortable with concepts and technologies used in materials focused on statistics. To address these concerns and to help inform dissemination efforts, the authors studied issues related to implementation of the statistics materials described here.

Framework

The teaching and learning of mathematics and statistics has been greatly influenced by the expanding capability and availability of technology (Ben-Zvi, 2000; Chance, Ben-Zvi, Garfield, & Medina, 2007; Heid & Blume, 2008). Others (Koehler & Mishra, 2005; Mishra & Koehler, 2008; Niess, 2005) have also considered how technology influences teaching and learning and have described a combination of technological, pedagogical, and content knowledge as a type of knowledge needed to effectively use technology to teach specific content. In considering a framework for the design and development of the first module on teaching statistics with technology, Lee and Hollebrands (2008b) specialized the notion of technological, pedagogical, and content knowledge for statistics. Figure 1 depicts teachers' technological, pedagogical, and statistical knowledge as layered circles with the foundation focused on teachers' statistical knowledge. The innermost layer, consisting of elements of technological, pedagogical, and statistical knowledge, is founded on and developed with teachers' knowledge in the outer two circles. Developing knowledge in the outer two layers of statistical knowledge and technological statistical knowledge is essential, but not sufficient, for teachers to develop the specialized knowledge represented in the innermost layer. The elements noted in each layer of Figure 1 are descriptors of the major foci of teachers' knowledge, thinking, skills and dispositions that the curriculum authors (Lee, Hollebrands, & Wilson, 2010) aimed to develop.

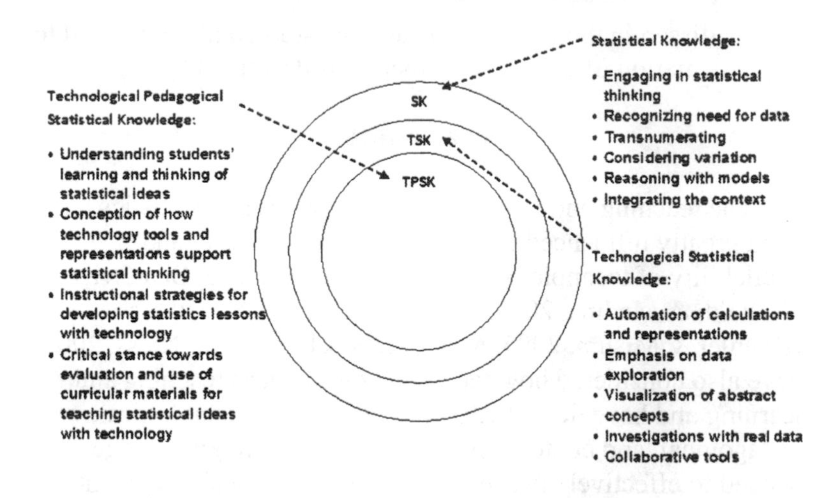

Figure 1. Framework for developing teachers' technological, pedagogical, and statistical knowledge.

For many teachers, engaging in statistical thinking is a different process than the processes involved in teaching and learning mathematics (del Mas, 2004). Thus, the statistics module materials engage teachers as learners and doers of statistics through exploring data of likely interest (e.g., national school data, vehicle fuel economy, birth data) and promote the practice of asking questions from data, while the technology facilitates use of various representations to analyze data in novel ways that motivate other questions to explore with the data.

The materials examine distributions graphically and characterize the data before computing statistical measures. Questions promote the comparison of distributions as a means of transitioning to thinking about data as aggregate and help teachers conceptually coordinate center and spread. Studying data in a univariate context helps students consider measures of variation (e.g., residuals, sum of squares) in a bivariate context when modeling with a least squares line. Throughout the materials, an aim is to develop teachers' understanding that the way one thinks in statistical contexts is often different than in pure mathematical contexts. The materials support consideration of how statistics is a tool for answering questions, but also that answers connected to the context of data are rarely strong enough to make definitive statements.

Technology-based tasks can simultaneously develop teachers' understanding of statistical ideas with essential technology skills and foster use of technology that amplifies or reorganizes one's statistical work (Ben-Zvi, 2000; Pea, 1987). The idea of an amplifier is that the tool expedites a process that could be completed without its use (e.g., automating computations, generating large lists of pseudorandom numbers, generating graphical representations). Technology tools can also be seen as reorganizers through such dynamic features as dragging, linking multiple representations, performing simulations, accessing large data sets, and using collaborative tools such as Wikis.

Throughout the statistics module, findings from research on students' understanding of statistical ideas are used to make points, raise issues, and pose questions for teachers. After teachers have engaged in examining a statistical question with technology, they are asked pedagogical questions concerning how technology and various representations can support or hinder students' statistical thinking.

Research Questions and Methods

This research focused on mathematics teacher educators' planning and debriefing when using the new curricular materials (Lee, et al., 2010) in a "Teaching Mathematics with Technology" methods course for middle and secondary prospective teachers. The research questions for this study are:

- How do mathematics teacher educators make sense of the intended curriculum and make decisions for implementation?

- In what ways do mathematics teacher educators use components of technological, pedagogical, and statistical knowledge in their planning and implementation of the curriculum?

Qualitative methods were employed to examine the local phenomena of planning and implementation through document collection and non-participant observer techniques (Huberman & Miles, 2002).

Context of Study

The study took place in Fall 2008 during a five-week unit on data analysis and probability in the methods course, which served prospective middle and high school teachers (hereafter referred to as "students"). The statistics module was used as the primary textbook during the unit. All students purchased the module with a CD containing necessary technology files. For this study, data from two chapters, Chapters 3 and 4, were analyzed (see Lee, Hollebrands & Wilson, 2010). These chapters make exclusive use of the software *Fathom* (Version 2.1), which limited the scope of the study to one form of technology. Both chapters are of significant length, with Chapter 4 conceptually building upon Chapter 3.

Participants

The course was taught by a mathematics teacher educator and a teaching assistant. The mathematics teacher educator had taught the methods course five times since Fall 2005, had seven years experience teaching middle and high school, and had implemented early field test versions of the statistics module. The teaching assistant had seven years experience teaching high school and two years at a community college. To facilitate an apprenticeship model, the pair met two hours weekly for planning and debriefing sessions. Because the focus was not on the instructors as individuals, they are collectively referred to as "teachers" and treated as a single unit of analysis.

Data Collection

The sources of data included audio recordings of the planning and debriefing sessions, and a copy of the teacher textbooks containing notes used during planning and instruction. One co-author attended the sessions as a non-participant observer, but occasionally asked a clarification question or interjected a comment that led to further discussion. A purposeful choice, however, was made not to observe the teaching directly so as not to introduce any anxiety about teaching performance. The use of planning and debriefing sessions as a source of data was also applied in the collegiate setting by Bartlo, Larsen, and Lockwood (2008) and was shown to effectively provide information about challenges instructors faced as well as a record of instructional decisions.

Analysis Methods

The transcripts of the planning and debriefing meetings (two sessions, four hours total) were analyzed to identify key themes about the teachers' decisions and use of resources. Transcripts were then coded for instances where it appeared the teachers drew upon elements of technological, pedagogical, and statistical knowledge. The codes were discussed until consensus was reached. Finally, examples that indicated where the teachers drew upon statistical knowledge, technological statistical knowledge, and technological pedagogical statistical knowledge (indicated in Figure 1) were sought.

Results

The coding process generated several common themes concerning how the teachers intended and used the curriculum. The results are organized by each research question.

Question 1: How do teachers make sense of the intended curriculum and make decisions for implementation?

The teachers used both the main student textbook materials and the technology files associated with Chapters 3 and 4. During the planning sessions, they discussed specific language (e.g., predictor/response), definitions (e.g., coefficient of determination), and conceptual descriptions (e.g., standard deviation) present in the chapters. Some ideas were unfamiliar or difficult for either them or, they hypothesized, for the students. For example, one of the teachers noted the terms univariate and bivariate in the text. The other said, "Yeah, especially that last—

that last paragraph in the intro, the difference between univariate and bivariate, 'cause some of them don't even know it." In addition, the teachers made several references to technical directions for how to perform an action in *Fathom*. It was evident that they opened and familiarized themselves with associated technology files. They knew the content of the files and even made a suggestion for additional helpful files. These examples illustrate that the teachers used the textbook and technology files as key faculty resources during planning and implementation. In contrast, there was little evidence of use of a detailed answer key during the planning discussions regarding answers to mathematical or pedagogical questions.

The majority of the teachers' time was spent discussing details of how to implement materials in the allotted class sessions. They discussed: (a) how much time they should spend on each section; (b) which sections they wanted to "cover" during class and which to assign for homework; (c) when they would need to access certain technology files during a lesson; and (d) when they should have students reference the textbook during a lesson. There were frequent discussions regarding facilitating whole class format versus small group work. The teachers tended to assign small group work for exploring data with technology where students were to use the textbook as a guide, and reserved pedagogy questions for whole group discussions because, "This is a teaching math with technology class, I like to go back and wrap up with the pedagogy questions."

Question 2: In what ways do teachers use components of technological, pedagogical, and statistical knowledge in their planning and implementation of the curriculum?

In coding the teachers' discussions, instances in their conversations were identified where at least one of them drew upon his or her understanding of one or more elements in the technological, pedagogical, and statistical knowledge framework. The knowledge of an individual was not described, but instead characterized by ways in which he or she may have been using elements of technological, pedagogical, and statistical knowledge in planning or implementation.

Statistical knowledge. In their discussions, the teachers exhibited an understanding of the importance of having students ask questions from data to support exploratory data analysis. The teachers also exhibited attention to center and spread in distributions as they discussed the importance of students' understanding of these ideas to make sense of the correlation coefficient and R^2. They further attended to how they could use univariate distributions displayed in box plots to ask students to conjecture how two univariate data sets could be displayed for a bivariate analysis using scatter plots.

Technological statistical knowledge. The teachers displayed an understanding of how to use technology as both an amplifier as well as a reorganizer (Ben-Zvi, 2000). They often referred to the ease of using technology to quickly create data displays such as box plots, allowing more time to discuss comparisons across different representations. They capitalized on technology as an amplifier as evidenced by their planning for and reflection upon class discussions regarding how representations highlight or mask data with regards to center and spread, and the effects of viewing data as individual points or in an aggregate.

The teachers also displayed their understanding of how technology could be used as a reorganizer for statistical concepts. They referenced benefits of the ability to use sliders in *Fathom* to explore the connections between the value of a correlation coefficient and the spread of data in a scatter plot. They frequently discussed how to highlight dynamically changing one representation and observing the effects in a linked representation. They indicated a goal for the students to take advantage of linked representations for their own learning and future teaching.

Technological pedagogical statistical knowledge. There were several instances when the teachers discussed the importance of teaching statistics conceptually, and how this may be difficult for students since many had learned statistics in a "formula-driven way." In these discussions, the teachers referenced diagrams in the textbook (e.g., diagrams for understanding deviations from a mean and residuals from line of best fit) and dynamic dragging capabilities in *Fathom* such as an ability to move data points and visualize the effect of the mean and deviations, as well as moving a line of best fit and the effect on associated squared residuals. They also discussed decisions to present material in a way different from what was explicitly suggested in the text. These instances illustrated a critical stance towards the curriculum and an ability to draw upon technological statistical knowledge and statistical knowledge to make planned or impromptu changes to the curriculum.

An example of a planned change to the curriculum occurred when the teachers discussed how to alter what was suggested in the module. Figure 2 shows the box plots for city and highway mpg that appeared in the module. The goal of the activity as presented in the materials was to begin with two univariate distributions and build to how the two attributes co-vary and can be represented in a scatter plot.

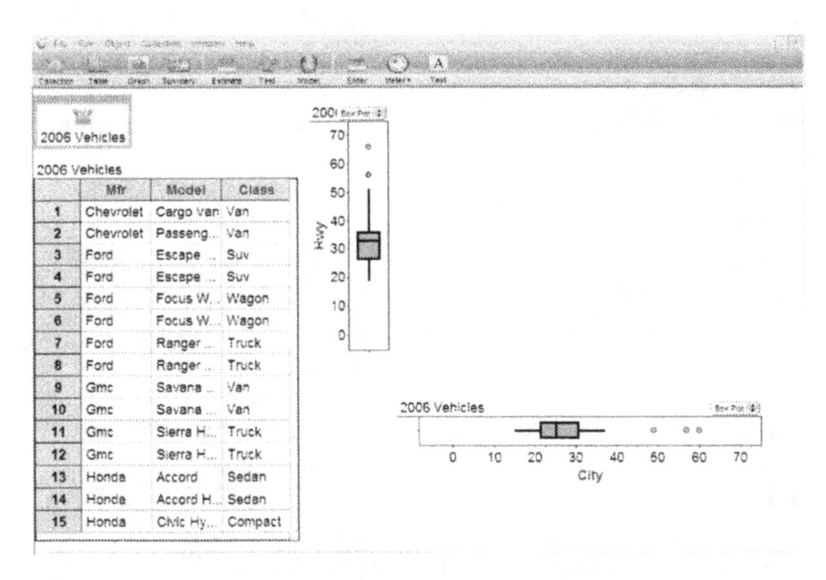

Figure 2. Graph of two box plots and space for a scatter plot.

Teacher 1: There's no question that asks them to explicitly —it might be nice to…say "Okay, based on your box plots; what kind of scatter plot do you anticipate?"

Teacher 2: The others [in the past], I think they viewed it more like an exercise: "We'll do this because that's what the book says" as opposed to making any connections as to why we were doing it. But yeah, I think that would be a good question.

Additionally, one of the teachers commented on how she "pointed out [during the lesson] some things about the cluster being where you would expect the interquartile ranges to intersect." In the module there is no mention of the interquartile range in this section (although it is discussed in other chapters). Thus she drew upon previous experiences with box plots and anticipated what the students knew about interquartile range and where that cluster should be in the scatter plot. This example shows how these teachers drew upon their technological, pedagogical, and statistical knowledge to notice that anticipating a scatter plot rather than simply following along and being shown a scatter plot could be a more effective way for students to understand connections between representations and the relationships between two attributes.

An example of an impromptu change to the curriculum occurred after one of the teachers noted that the students in this course had little experience with residuals and the resulting understanding was somewhat limited. In a problem in the statistics module, students were asked to: (a) create a residual plot, (b) adjust the moveable line, (c) consider the residual plot when determining the usefulness of a linear model, (d) sketch the location of the moveable line given a residual plot, and (e) describe some of the conceptual difficulties high school learners may have and suggest ways to aid their understanding. The teachers noticed that the students had many unresolved questions regarding residuals.

One of the teachers reflected upon how she decided to use a *Fathom* technology resource file for a Chapter 4 lesson to provide a demonstration of the residual plot and how it relates to a moveable line. She used vertical translations of the moveable line to illustrate how the residual plot would respond. As shown in Figure 3, the teacher placed the moveable line entirely above all of the data points, to help students understand why the corresponding residuals would have a negative numerical value. A similar translation was performed to exemplify all positive-valued residuals. This drastic manipulation of the moveable line further augmented the relationship between the moveable line and residual plot.

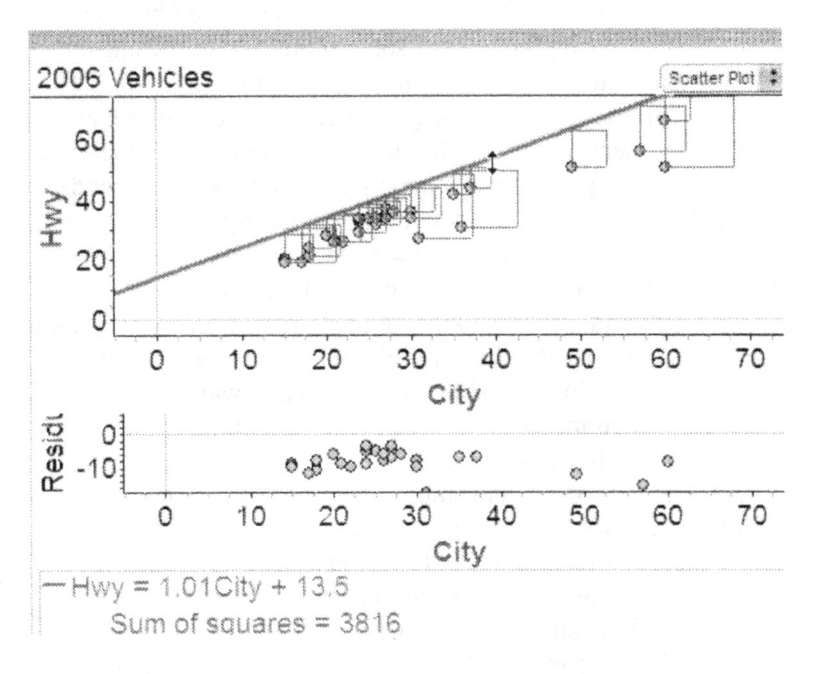

Figure 3. Dynamically linked scatter plot, moveable line, squared residuals, and residual plot.

Upon reflection, the teacher noted that following her demonstration students were able to successfully respond to questions that were previously incomplete. In this example, the dynamic environment was used as a reorganizer to reinforce the connections between the scatter plot, moveable line, and residual plot, illustrating an instructional decision that was made for the development of statistical understanding through the use of technology. The teachers were able to conceptualize how the dynamic representations supported students' statistical thinking.

Discussion

An analysis of the planning and debriefing conducted by the teachers when implementing two chapters of the statistics module helped the authors understand ways that teachers may make sense of the materials and the major decisions made when implementing the module. It became clear that teacher resource materials needed to include suggestions for how to allocate time for each section in the module as well as ways to utilize small group work and whole group discussions. By discovering what decisions teachers make when implementing a new curriculum, the authors can revise the teacher resource materials to address issues and difficulties encountered by those first implementing the materials.

For example, since residual plots are likely a new representation for many students, perhaps the textbook section on visualizing residuals is best addressed as a whole class discussion. Further revisions should include more descriptions of a residual plot and more explicit connections to materials presented in Chapter 3 on deviations from a mean, which lays a conceptual foundation for residuals from a line of best fit. Results such as these will influence future professional development for faculty. One idea for professional development is to use examples of how faculty implemented a particular chapter, leading to discussions of pros or cons faculty can consider for their own implementations.

Results also indicated that teachers needed to draw upon the types of knowledge represented in the technological, pedagogical, and statistical knowledge framework (Figure 1). Not only did this help validate the framework as being useful for characterizing elements of technological, pedagogical, and statistical knowledge, it can help in developing an instructor's guide that makes the framework explicit to teachers. One interesting finding was that instances of technological, pedagogical, and statistical knowledge found in the analysis seemed to occur when a teacher extended the written curriculum to improve the understanding of prospective teachers. In addition, it appears that in instances where they created learning opportunities for students, teachers needed to draw upon several elements of technological, pedagogical, and statistical knowledge:

1. Understanding students' learning and thinking of statistical ideas with and without technology.
2. Conception of how technology tools and representations support statistical thinking.
3. Instructional strategies for developing statistics lessons with technology.
4. Critical stance towards evaluation and use of materials for teaching statistical ideas with technology.

In the box plot example, the teachers used elements 2, 3, and 4. The residual plot example included elements 1, 2, and 4. In future research, the authors will examine whether this generalizes to how and when other teachers draw upon technological, pedagogical, and statistical knowledge.

The framework presented here describes the intended ways the curriculum materials were designed to develop this type of knowledge for students. Ultimately students who learn from these materials will develop their own technological, pedagogical, and statistical knowledge that can be useful in teaching statistics to future learners in grades 6–12. However, the results of this study suggest that more ways are needed to develop technological, pedagogical, and statistical knowledge for mathematics teacher educators as they work towards developing similar knowledge in their prospective and practicing teachers.

References

Association of Mathematics Teacher Educators. (2006). *Preparing teachers to use technology to enhance the learning of mathematics.* Retrieved from http://www.amte.

net/sites/all/themes/amte/resources/AMTETechnologyPositionStatement.pdf.

Bartlo, J., Larsen, S., & Lockwood, E. (2008). *Scaling up instructional activities: Lessons learned from a collaboration between a mathematician and a mathematics education researcher.* Paper presented at the Conference on Research in Undergraduate Mathematics Education. San Diego, CA. Retrieved from http://mathed.asu.edu/crume2008/Proceedings/Bartlo%20LONG.pdf

Ben-Zvi, D. (2000). Toward understanding the role of technological tools in statistical learning. *Mathematical Thinking and Learning, 2*(1 & 2), 127–155.

Chance, B., Ben-Zvi, D., Garfield, J., & Medina, E. (2007). The role of technology in improving student learning of statistics. *Technology Innovations in Statistics Education 1*(1). Retrieved from http://www.escholarship.org/uc/item/8sd2t4rr

College Board. (2007). *AP Program Size and Increments*. Retrieved from http://apcentral.collegeboard.com/apc/public/repository/2007_Size_and_Increment.pdf

del Mas, R. C. (2004). A comparison of mathematical and statistical reasoning. In D. Ben-Zvi & J. Garfield (Eds.), *The challenge of developing statistical literacy, reasoning, and thinking* (pp. 79–95). Netherlands: Kluwer Academic Publishers.

Franklin, C., Kader, G., Mewborn, D., Moreno, J., Peck, R., Perry, M., & Scheaffer, R. (2005). *Guidelines for assessment and instruction in statistics education (GAISE) report: A pre-k–12 curriculum framework.* Alexandria, VA: American Statistical Association.

Heid, M. K., & Blume, G. (2008). *Research on technology and the teaching and learning of mathematics: Volume 1.* Charlotte, NC: Information Age Publishers.

Huberman, A. M., & Miles, M. B. (2002). *The qualitative researcher's companion.* Thousand Oaks, CA: Sage.

Key Curriculum Press. (2010). *Fathom* 2.1. Emeryville, CA: Author.

Koehler, M. J., & Mishra, P. (2005). What happens when teachers design educational technology? The development of technological pedagogical content knowledge. *Journal of Educational Computing Research, 32*, 131–152.

Lee, H. S., & Hollebrands, K. (2008a). Preparing to teach mathematics with technology: An integrated approach to developing technological pedagogical content knowledge. *Contemporary Issues in Technology and Teacher Education [Online serial], 8*. Retrieved from http://www.citejournal.org/vol8/iss4/mathematics/article1.cfm

Lee, H. S., & Hollebrands, K. F. (2008b). Preparing to teach data analysis and probability with technology. *Proceedings of the Joint Study of the ICMI/IASE*. Monterrey, MX, June. Retrieved from http://www.ugr.es/~icmi/iase_study/Files/Topic3/T3P4_Lee.pdf

Lee, H. S., Hollebrands, K. F., & Wilson, P. H. (2010). *Preparing to teach mathematics with technology: An integrated approach to data analysis and probability (1st ed.)*. Dubuque, IA: Kendall Hunt Publishing.

Mishra, P., & Koehler, M. (2008, March). *Introducing technological pedagogical content knowledge*. Paper presented at the Annual Meeting of the American Educational Research Association, New York.

National Council of Teachers of Mathematics. (2000). *Principles and standards for school mathematics*. Reston, VA: Author.

Niess, M. L. (2005). Preparing teachers to teach science and mathematics with technology: Developing a technology pedagogical content knowledge. *Teaching and Teacher Education, 21*, 509–523.

Pea, R. D. (1987). Cognitive technologies for mathematics education. In A. Schoenfeld (Ed.), *Cognitive science and mathematics education* (pp. 89–122). Hillsdale, NJ: Erlbaum.

Hollylynne Stohl Lee is an associate professor of mathematics education at North Carolina State University. She focuses her work on understanding the teaching and learning of statistics and probability, particularly with dynamic software tools. She can be reached at hollylynne@ncsu.edu.

Sarah Ives is an assistant professor of mathematics at Texas A&M University–Corpus Christi. Her research is focused on the use of artifacts of practice in teacher education, specifically in teaching probability and statistics. She can be reached at sarah.ives@tamucc.edu

Tina Starling is a third year doctoral student in mathematics education at North Carolina State University. Her research interests include teaching mathematics with technology and teaching and learning mathematics education online. Her email address is ttstarli@ncsu.edu

Karen Hollebrands is an associate professor of mathematics education at North Carolina State University. Her research focuses on issues related to the teaching and learning of mathematics with technology. She can be contacted at karen_hollebrands@ncsu.edu

Lenges, A.
AMTE Monograph 7
Mathematics Teaching: Putting Research into Practice at All Levels
© 2010, pp. 25–40

3

Enhancing Specialized Content Knowledge for Teaching Mathematics through Authentic Tasks of Teaching in a Professional Learning Environment

Anita Lenges
The Evergreen State College

The mathematics education community generally agrees that deep and robust mathematics knowledge supports teachers' pedagogical decisions thus enhancing student learning. Teacher educators with this perspective focus on specialized mathematical content knowledge for teaching mathematics (Ball, Thames, & Phelps, 2008). As they address this focus, questions arise about how to advance this specialized knowledge, how to help teachers value this knowledge, and how to deepen the knowledge so that it is useful in teaching practice. One effective way to enhance teachers' knowledge in professional learning settings is through collaborative work on lesson planning. This approach may help teachers revise their everyday approaches to lesson planning as they more deeply consider their own mathematics knowledge. However, using pedagogy as a vehicle for learning mathematics may require facilitation to nudge mathematical thinking to the fore.

Mathematics teacher educators are learning that teaching demands a form of mathematical knowledge identified as specialized content knowledge (Ball et al., 2008). This specialized knowledge is purely mathematical and uniquely used in the work of mathematics teaching. For example, while students commonly solve computation problems, a skilled teacher is able to examine a student's strategy for solving a problem to determine where the work might be incorrect as well as what the student understands and what the student has yet to learn. Beyond that, he or she can examine a student-invented strategy to determine if the strategy generalizes to all cases within a class of problems.

This mathematical work is specialized in the field of teaching. Yet teacher educators wonder how to help teachers develop such knowledge. Many teacher educators attempt to enhance teachers' mathematical knowledge through engaging them in rich mathematical tasks (e.g., analyzing the border of an m x n swimming pool or the number of dots in a growing pattern). Others use tasks of teaching, such as examining students' strategies for solving problems (Schifter, Bastable, & Russell, 1999). Purely mathematical explorations allow teachers to expand their mathematical knowledge in professional learning settings. Teachers may also need to develop their mathematical knowledge during lesson planning if their mathematical knowledge is limited. Embedding mathematical explorations in pedagogical practice in professional learning situations may help teachers normalize that as routine. This paper describes an experience where goal setting and lesson planning were used to activate and enhance teachers' specialized mathematics content knowledge.

When helping teachers expand their mathematical knowledge through solving complex mathematics problems, teacher educators may expect that in-service teachers automatically use their newly gained mathematical knowledge with students in their own classrooms. Teacher educators model effective teaching practices and reveal their pedagogies for teaching the particular content ideas of a task through experience, reflection and discussion. Some explicitly tie the content to student work where students did the same or a similar task. Teacher educators may imagine that the teachers are now more able to help their students develop a deep and robust mathematical knowledge similar to what they have experienced in professional development.

However, it is unclear whether purely mathematical explorations for in-service teachers are likely to lead to instructional improvement, particularly when teachers do not teach their students using the same tasks on which they work with colleagues in their own professional development. Another major challenge to using purely mathematical explorations in professional development is the difficulty in motivating all teachers to delve deeply into the mathematics. Some teachers may feel they already know the content, or they may not see the relevance of the mathematical inquiry for practice.

An alternate approach to enhancing teachers' mathematical knowledge is through teaching practices, where mathematical explorations are inserted into common aspects of a teacher's work. Research on teacher learning indicates the importance of keeping teacher learning opportunities close to practice and centered on important professional activity (Ball & Cohen, 1999; Grossman,et al., 2009). Some professional development materials reflect this design. Teachers solve mathematics problems and examine associated student work (Carpenter, Fennema, Franke, Levi, & Empson, 1999; Schifter, 1998), and reflect on cases of teaching and learning (Schifter et al., 1999; Smith, Silver & Stein, 2005). Facilitation maintains explicit focus on the mathematics. Experience shows that teachers' mathematical knowledge can, indeed, be enhanced through professional development that incorporates teaching practices (Bell, Wilson, Higgins, & McCoach, 2008).

When professional development is structured to engage teachers in teaching practices centered on worthwhile mathematical explorations, it creates an opportunity for teachers to enhance their mathematics knowledge and revise teaching practices to demonstrate this knowledge. With this approach, teachers may be more likely to recognize the intent of the explorations as enhancing their teaching skills, thus motivating their work (Bransford, Brown, & Cocking, 1999). Such simulations of teachers' work in professional development can be rehearsals of real teaching experiences with the benefits of collaboration, reflection and potential enhancements to teachers' practice.

This paper shows an example of deepening teachers' mathematical knowledge through practice-based professional development—in this case, lesson preparation. It also provides several specific examples of how the teacher educator's actions may influence whether or not specialized content knowledge is engaged and enhanced for teachers.

Introduction to the Activity

The following sections outline an activity used by in-service teachers in a Master of Education program for elementary, middle, and high school teachers seeking an endorsement in secondary or middle-level mathematics. The activity was part of a course titled "Teaching Math for Social Justice." The class of 11 teachers and one instructor formed a professional community engaged in critical colleagueship (Lord, 1994) over three academic quarters. Teacher-learners were provided with the experience of modifying a broadly written unit plan and changing it into a rich mathematical task embedded in a problem context appropriate for the middle grades. The original unit plan, "Create the Average Person," invited middle school students to create an average person on earth. Students explored humans along a broad set of characteristics such as age, eye-color, gender, religion, interests, career, living conditions, income, etc. The intent of the task was to de-center American dominant culture and whiteness while teaching the statistical concepts of mean, median and mode.

The instructor wanted the in-service teachers to understand these statistical concepts beyond a procedural knowledge level. She wanted them to exhibit improvement of teaching effectiveness through a clear and precise understanding of these measures. Finally, the instructor wanted teacher-learners to recognize the importance of understanding the mathematics in preparing a strong lesson. The instructor believed that aligning learning goals with a given task or constructing a task to advance particular learning goals would propel teachers into deeper mathematics.

In order to insert the mathematical exploration into teachers' work and to prepare a strong lesson to support student learning, the instructor assigned teacher-learners to clearly name the mathematical goals of the task and identify mathematical competencies students would bring to the task. In order to effectively complete the task requirements, teacher-learners necessarily had to explain the mathematics involved and articulate their mathematical ideas clearly to each other.

The teacher-learners were encouraged to establish a cognitively demanding (Stein, Smith, Henningsen, & Silver, 2000) problem context (Fosnot & Dolk, 2002) to develop substantive mathematical knowledge rather than simply a context for applying previously learned knowledge. The problem context was to provide relevance for middle grades students as they explored the statistical concepts grounded in real life. For example, after discovering that brown is the most prevalent eye color, the students might realize that reporting only the most dominant eye color in the construction of the average person eliminated all other colors, thus deleting demographic data about the population in general. Or students might come to see that reporting only a median income gives no sense of the range (or spread) of the data. These are important mathematical insights that may be lost in a curriculum focused primarily on procedures for finding mean, median, and mode.

Writing Goals as a Mathematical Experience

As the in-service teachers worked on this task in groups of three or four, some moved directly into naming the goal and writing the task for students, bypassing a significant examination of the mathematics. Others struggled to define the task to be given to students. As the teacher educator heard or saw groups' vague goals, she asked them to be precise in naming "specifically what you want students to know." She urged the class to state goals clearly while she provided comparisons of vague and specific goals. At the end of class that session, one group of teachers produced the mathematical goals shown in Figure 1.

Goals for "Find the Average Person"

1. Students will be able to calculate the mean, median and mode for a given data set.
2. Students will be able to understand the meaning of each statistical measure: mean, median, and mode.
3. Students will be able to understand that some data may be described more effectively using one measure than another.
4. The mean is affected by outliers or extreme data.
5. The median is less affected by outliers or extreme data.
6. The mode is the most common number of a data set. There could be multiple modes.

Figure 1: One group's goals for "Find the Average Person" project

The instructor felt that Goals 1, 4, and 5 were sufficiently precise. A teacher who knows the mathematics sufficiently could recognize when a student mastered these ideas or skills. The instructor and the group also recognized the need for greater precision in Goals 2 and 3. Their vagueness made it unclear how a teacher would know if a student had achieved those goals, and not every teacher in the group would necessarily agree on when mastery was achieved. The group was also content with Goal 6, despite the fact that some statistics texts claim that a data set might not have a mode.

One Group's Story

One week after the lesson, the instructor interviewed the group of three teachers who wrote these goals. They elaborated on their experience and how they arrived at a higher degree of specificity in stating mathematical goals four and five. These teachers were particularly reflective, collaborative and willing to work hard to understand fundamental mathematical ideas and pedagogies. They also brought significant knowledge, experience and resources to the discussion. Judith, an elementary school teacher with 15 years of teaching experience, had participated in many mathematics professional development seminars in her career. A facilitator for Developing Mathematical Ideas seminars, she was seeking her second master's degree. Raymond had been both an elementary school teacher and a school counselor, working with children for 13 years. Marilyn directed a child care center at a college, and had been a substitute teacher for 11 years. She was also seeking her second master's degree. All three teachers were additionally seeking a secondary mathematics endorsement. Results of the group's collaborative work are discussed in the following sections.

Doing the Mathematics as Learners

Following the instructor's recommendation and their own inclination, this group of teacher-learners began the task as if they were middle school students. "We needed to get an idea of what the task was asking. Just doing it was helpful." To consider what characteristics students might study about humans in the world, they listed characteristics of themselves, and generalized those to what categories they represented. For example, Marilyn listed "Blue eyes" as eye color, "Grey hair" as hair color, "Hike" as interests, "Love chocolate" as food preferences, and "Come from low-income family" as family income. Her work is shown in Figure 2.

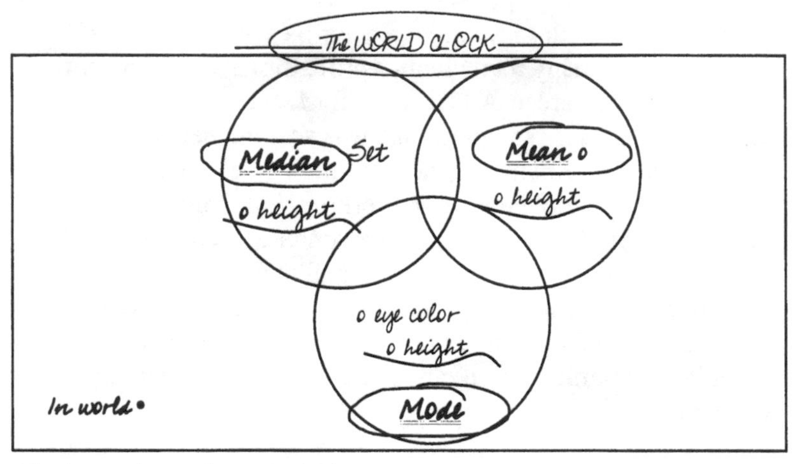

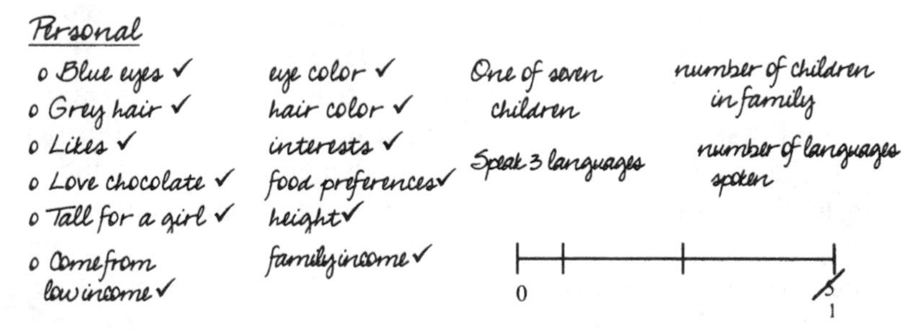

Figure 2. Marilyn's written class work

Seeking an Overarching Theory of Mean, Median and Mode

As they worked, the three teachers developed questions about mean, median, and mode and the circumstances under which they are useful and appropriate measures. They also thought about particular questions that would be best answered with one or another of these measures. They pursued this question in detail, following the instructor's suggestion that Goal 3 was too vague and that they needed to be able to agree on mastery among themselves before they could require that students learn this key idea.

Judith: We were trying to think of a way that kids could begin to take those characteristics and attach [them] to the idea of mean, median and mode. And I got the idea of this three-part Venn diagram, each circle representing one of the measures of central tendency. Then we got into quite the discussion. We were talking about each characteristic.... We kind of had this naive idea that this one ... would be mean ... median would be the best way to measure this. Then you started realizing that there were certain ones that could only be measured by mode, like eye-color. And there were certain things that could be measured in more than one category, like height. It could be mean or median.

Marilyn: Or mode.

Judith: We put height in the middle. (Judith's diagram, not shown here, has height in the intersection of all three circles.)

The teacher-learners recognized that they had to make sense of mathematical questions that could surface for students while doing this activity. If they gave their students this "Create the Average Person" task, they would need to be very familiar with the mathematical terrain in order to make decisions about the task and to focus dialogue toward their goals.

The group then entered a conversation about gender, and whether it represented categorical data with two categories (male and female) or required a third category to include transgender, in either case leading to the use of mode. They also considered a continuum that could have decimal values and be described by measures of mean or median. Their discussion is shown below. The question of using a continuum surfaced because they realized that in order to use mean or median, they would need numerical data.

Marilyn: We kept coming to gender and we were trying to sort out gender…like would it fit in this or this. Like, gender, median? You can't do that. Gender, mean?

Anita: Why was gender so interesting in that way?

Marilyn: Because it was only a two-choice thing.

The group then described trying to assign number values to gender, such as 0 for boys and 1 for girls.

Judith: But then we got into the whole conversation of if there [are] only two choices in terms of gender.

Marilyn: Like maybe match it up [on a scale of 1] to 5. And then how could you put a number on that, like a mean. That was the one that crystallized for me that there is *definitely* [verbal emphasis] some sort of difference between mean, median, and mode…how could you say that the average person is 4.6 male? It is not sort of a sensible statement.[1]

This is an issue for students when determining an average person. The key mathematics the teachers collectively named in the conversation was that there exist categorical data (a term the instructor provided for them in discussion), and that categorical data can easily be described with mode. They also grappled with the possibilities of representing gender on a line and how that related to mean, median, and mode. A subsequent discussion of eye color further solidified the group's understanding of categorical data. Finally, the teacher-learners developed the idea of mutually exclusive categories.

---

[1] Current understanding of biological sex, gender identity, and gender expression demonstrate that a continuum could hold more meaning than gender or sex categories.

Further Clarifying Goal 3

The group continued to share their prior knowledge and make sense of each other's knowledge in light of their developing understanding. Marilyn raised the idea that housing costs are reported by the median. Judith, Marilyn and Raymond made collective sense of how a single expensive house would have a more significant affect on the mean than a single inexpensive house.

Judith: I was trying to figure out why the median [is used to describe housing prices] … poor person's house, rich person's house, poor person's house, rich person's house (using her hands to demonstrate counting in from the extremes to find the middle-most house cost.). And so, it's more accurate than if you were to average them.… A house has got to cost a bit, you know what I mean, there are no houses at $500. So there is a point at which poorer people's houses and middle class people's houses are in common.…

Marilyn: If you put one of your bazillion dollar houses and you average them with ours, it would look like we have really expensive houses.… Because it would throw it way off the mean.

During the second class session, Marilyn shared information with her group that she gathered from an old statistics text with the group that further emphasized how the mean is strongly influenced by extreme data. Judith shared how she taught her elementary students about mean as combining all quantities, such as the number of pets in each home, and redistributing them equally among the students in her class. These discussions further clarified their understanding "that the mean is more affected by outliers, and the median is less affected by outliers."

Preparing a Cognitively Demanding Problem Context for Students

As the group worked to articulate the task for students, they wanted to stack the mathematical deck to ensure that particular mathematics would surface, causing students to grapple with the same issues they had. In particular, they wanted to make sure students encountered human characteristics best described by the mode, and characteristics that could be described by more than one measure, requiring them to determine and justify a particular choice. However, they also wanted to leave the task sufficiently open to motivate student engagement and creativity.

The group deliberated about requiring a Venn diagram, feeling that the diagram would force students to consider the possibility of overlapping spaces where two measures would be reasonable ways of describing a particular human characteristic. But would students automatically generate possibilities of overlapping categories without requiring the Venn diagram, or would they naively assume that there is only one way to describe any human characteristic, as the teacher-learners had done initially? In the end, the group chose to require the Venn diagram. Marilyn noted that "even if they didn't have one in [the intersection of all three circles], they would have to reconsider that something *could* be in all three." Judith added, "It was a way to get them to really deepen their thinking about those three things, not just different ways to calculate the same thing."

This debate, i.e., how to construct the task so that important mathematical ideas and dilemmas surface for students, relied on the teachers' own experiences and knowledge as they deliberated about the mathematics. While there was more mathematics for the group to clarify for this task to be well developed, the mathematical work they did was central to making a critical decision in designing the task. If, on the other hand, the group had moved right into the task of modifying the lesson without engaging in their own deep mathematical inquiry, they would not have been sufficiently prepared to construct a cognitively demanding problem context.

The last conversation illuminates where teachers' mathematical knowledge interacts with pedagogical decision-making in a crucial way. They now had a collective understanding about some concepts surrounding the statistical measures of mean, median and mode that they felt were important for students to learn. These went beyond simply computing measures. The group wanted students to relate the statistical measures to each other if possible, to consider conditions under which one measure might be preferable to another, and to recognize that some data sets (e.g., eye color) can be analyzed with only one measure, while other data (e.g., housing costs) lend themselves to multiple measures. Finally, they grappled with the value of the representation of a Venn diagram and its mathematical potential for student understanding.

Deepening Understanding by Becoming Mathematically Specific about Goals

The teachers in this group were asked both how and why they developed more specific goals about mean and median such as goals four and five. Judith articulated how they began to take up the idea of goal specificity and mathematical clarity. She explained that the group's effort to clarify the mathematics for themselves was motivated by the instructor asking for an articulation of clear mathematical goals.

Judith: There was a point at which you pressed us.... We might have had that first goal ("Students will be able to calculate the mean, median and mode for a given set of data") and I think you wanted ... what do we really want them to understand? I think we might have come up with this goal ("Students will be able to understand the meaning of each measure: mean, median, and mode") but ... what do we really want them to *understand* [verbal emphasis] about each measure?... We realized that it is one thing to be able to calculate them and know how you get them, and it is another thing to really understand it. And that came out of this discussion of the time before, how important it is to really understand the *meaning* [verbal emphasis] of each and how each of those might be affected by a particular set of data.

Raymond: We kind of had it in our heads, but we didn't really articulate it until you pushed us on it. What is it about the mean? What is it about the median? And that is where we kind of hit a wall for a little bit before we got through it…. But I think it was because of your pressing that caused us to break each one of those down a little bit more, and [ask] "What is it about the mean? What is it about the median?" Because … I think we had this central tendency kind of a catch-all….

Had the facilitator not required goal specificity, teachers may not have become as clear about the mathematics in the activity. This highlights the importance of facilitation to advance the mathematical work when using practices of teaching as the venue for mathematical development.

Conclusion

In this exploration, teachers moved between mathematical and pedagogical work, largely due to the framing of a professional learning experience involving lesson planning and goal setting. Particular task requirements and facilitation emphasized mathematical specificity and clarity that supported a deep consideration of the mathematics. Specialized content knowledge emerged as the need arose to understand the mathematics for teaching, to write clear mathematical goals, and to scaffold the task so students bump up against important mathematical dilemmas. The suggestion that normalizing deeper mathematical exploration as a regular part of lesson preparation (the desired effect) is likely to occur with greater ease when critical elements of an authentic situation (teaching practice) are present in the professional learning setting.

References

Ball, D. L., & Cohen, D. (1999). Developing practice, developing practitioners: Toward a practice-based theory of professional education. In J. Boaler (Ed.), *Multiple perspectives on mathematics teaching and learning* (pp. 83–104). Westport, CT: Ablex.

Ball, D. L., Thames, M. H., & Phelps, G. (2008). Content knowledge for teaching: What makes it special? *Journal of Teacher Education, 59*, 389–407.

Bell, C., Wilson, S. M., Higgins, T., & McCoach, D. B. (2008). Measuring the effects of professional development: The case of developing mathematical ideas. Retrieved from http://www.mathleadership.org/upload/docs/13_0.pdf

Bransford, J. D., Brown, A. L., & Cocking, R. R. (Eds.). (1999). *How people learn: Brain, mind, experience, and school.* Washington DC: National Academy Press.

Carpenter, T., Fennema, E., Franke, M., Levi, L., & Empson, S. (1999). *Children's mathematics: Cognitively guided instruction*. Portsmouth, NH: Heinemann.

Fosnot, C. T., & Dolk, M. (2002). *Young mathematicians at work: Constructing factions, decimals, and percents*. Portsmouth, NH: Heinemann.

Grossman, P., Compton, C., Igra, D., Ronfeldt, M., Shahan, E., & Williamson, P. (2009). Teaching practice: A cross-professional perspective. *Teachers College Record, 111*, 2055–2100.

Lord, B. (1994). Teachers' professional development: critical colleagueship and the role of professional communities. In N. Cobb (Ed.), *The future of education: Perspectives on national standards in America* (pp. 175–203). New York: The College Board.

Schifter, D. (1998). Learning mathematics for teaching: From a teachers' seminar to the classroom. *Journal of Mathematics Teacher Education, 1*, 55–87.

Schifter, D., Bastable, V., & Russell, S. J. (1999). *Developing mathematical ideas: Building a system of tens, casebook and facilitator's guide, Volume 1*. Parsippany, NJ: Dale Seymour Publications.

Smith, M. S., Silver, E. A., & Stein, M. K., with Boston, M., Henningsen, M., & Hillen, A. (2005). *Improving instruction in rational numbers and proportionality: Using cases to transform mathematics teaching and learning, Volume 1*. New York: Teachers College Press.

Stein, M. K., Smith, M. S., Henningsen, M., & Silver, E. A. (2000). *Implementing standards-based mathematics instruction: A casebook for professional development*. Reston, VA: Teachers College Press.

Anita Lenges is a faculty member in the Teacher Education Programs at The Evergreen State College. Her work centers on anti-racist, anti-biased teacher preparation that supports students to thrive in mathematically rich environments.

Spangler, D. A.
AMTE Monograph 7
Mathematics Teaching: Putting Research into Practice at All Levels
© 2010, pp. 41–55

4

Relationships between Content Knowledge, Authority, Teaching Practice, and Reflection

Denise A. Spangler
University of Georgia

This paper presents a description of ways that elementary school teachers' mathematical content knowledge and locus of authority are related to their instructional practices and their reflections on their teaching. The cases of four teachers are used to illustrate four possible combinations of content knowledge (high/low) and authority (internal/external). Although four cases are presented as exemplars, teachers' practices were not consistent across time and setting, and the changes were not necessarily reflective of "growth" toward more reform-oriented teaching. Data were drawn from a five-year study in which two cohorts of elementary school teachers were followed from their junior year in college through their first two years of teaching.

In previous research, the author identified connections between the locus of authority (internal or external) from the viewpoint of preservice teachers and their propensities to think reflectively about pedagogical dilemmas (Mewborn, 1999). It was proposed that when preservice teachers see themselves as having the authority to raise and solve pedagogical dilemmas, they are more likely to think reflectively. In contrast, when preservice teachers see authority as external to them (residing in a teacher educator, an experienced classroom teacher, or in a textbook), they are inclined not to think reflectively about pedagogical dilemmas. The study reported here extends this connection to include the way that authority and mathematical content knowledge are interconnected in shaping teaching practice and reflection on that practice. In particular, this chapter describes four possible interactions between authority (internal/external) and mathematical content knowledge (high/low) and associates each with a particular approach to teaching mathematics and reflecting on that teaching.

Theoretical Perspective

This project is situated within the interpretive paradigm for teacher socialization (Zeichner & Gore, 1990). This interpretive approach involves an attempt to understand the nature of a social setting at the level of subjective experience. The purpose of this approach is to gain an understanding of the situation from the perspective of the participants and within their levels of consciousness and subjectivity. The goal is to "capture and share the understanding that participants in an educational encounter have of what they are teaching and learning" (Kilpatrick, 1988, p. 98). Eisenhart (1988) noted that the purpose of research questions posed by researchers using the interpretive paradigm is first to describe what is "going on" and second to uncover the "intersubjective meanings" (p. 103) that undergird what is going on in order to make them reasonable.

Methods

Data reported here were collected as part of a five-year research project entitled Learning to Teach Elementary Mathematics[i] in which two cohorts of preservice teachers were studied for two years of their preservice program and their first two years of teaching. The overall project goal was to develop conceptual frameworks for understanding teaching and learning in elementary mathematics teacher education by studying how novice teachers craft teaching practices across time as a result of personal theories, teaching experiences, and teacher education programs.

Project participants were selected from two cohort groups who began a four-semester teacher education program in the fall of 2000 (Group A) and 2001 (Group B). Some data were collected on all students from each cohort, but the majority of data collection focused on two target subsets—six students from Group A and nine students from Group B. The target students were selected by purposeful sampling (Bogdan & Biklen, 1992) to represent a range of personal theories about mathematics. To the extent possible, target students were reflective of the diversity of students enrolled in each cohort. The 15 target students consisted of 13 White women, one White man, and one African American man. Although the racial and gender composition of target students was fairly homogeneous, there was considerable diversity in their experiences with mathematics and their personal theories about the teaching and learning of mathematics.

The data set includes results from a mathematics beliefs survey (Ambrose, Phillip, Chauvot, & Clement, 2003), a content knowledge assessment (Hill & Ball, 2004), and all written work produced by the students during their two mathematics methods courses. In addition, target students were observed teaching a mathematics lesson four times and individually interviewed on four occasions during their preservice years. Observations occurred during field experiences in the second and third semesters of the program (one observation each) and during the student teaching experience (two observations). The four interviews were semi-structured and took place at the end of every semester of the teacher education program.

The first interview was conducted near the end of the first mathematics methods course and focused on eliciting autobiographical data from the participants regarding their views of mathematics and their prior experiences as mathematics learners. They were also asked to reflect on their experiences working one-on-one with a student in mathematics during the methods class. The second and third interviews occurred at the end of the second and third semesters of the teacher education program and asked participants to reflect on their practicum field experiences and how these experiences differed (or not) from their one-on-one teaching experiences and from what they had learned in the mathematics methods course. The fourth interview, conducted at the conclusion of student teaching, focused on reflections about various teaching experiences during the teacher education program and changes (if any) in the participants' conceptions about mathematics and mathematics teaching and learning from the beginning to the end of the teacher education program. Additionally, most target students (those teaching within reasonable driving distance) were observed teaching mathematics monthly and interviewed twice per year during their first two years of teaching.

All data were transcribed and organized for coding purposes, and the research team defined an initial set of 25 codes from the research literature. Line-by-line coding of data took place chronologically for all 15 target students, with different researchers coding data for different participants and then writing summaries, or "data stories." The chronological coding and parallel data stories facilitated comparison and contrast among participants. From this initial round of coding and team sharing of data stories, a final set of codes was developed. The data stories and original data were re-coded across participants using the new set of codes.

Results

The grid in Figure 1 depicts the potential relationships between content knowledge and view of authority. The description in each cell characterizes the nature of the corresponding lessons and post-lesson reflections as they were identified from data coding and analysis. This is followed by a series of descriptions of a teacher who typifies each cell, with an example of a lesson she taught depicting the type of teaching and reflection that is suggested by the category.

Authority		Content Knowledge	
		High	Low
Authority	Internal	Lessons are student-centered, mathematically rich, developmentally appropriate, and teacher is reflective about self, students, and content after the lesson.	Lessons focus on fun activities with some mathematical substance, but they sometimes fail due to the preservice teachers' lack of mathematical foresight or inability to "think on the fly." Teacher is reflective after the lesson, but reflections center mostly on self and teaching actions.
Authority	External	Lessons consist of attempts to provide clear and concise procedural explanations. Preservice teachers assume that this is how children learn mathematics best and that improving their teaching is merely a matter of giving better/ clearer explanations. Little reflection after the lesson.	Preservice teachers try to provide lessons that make math fun and easy in order to spare students the agony of learning. The content of the lesson is often unclear and sometimes mathematically inaccurate or unimportant. There is little reflection on the lesson beyond whether or not the students enjoyed it.

Figure 1. Relationship between content knowledge, authority, teaching practice, and reflection.

Jayne

Jayne is an example of a teacher with high mathematics content knowledge and an internal locus of authority. She described herself as favoring language arts and mathematics, with a particular affinity for algebra. She also described herself academically as "creative," "original," "independent," "wanting to stand out from the crowd," and "not afraid to ask for help." Her scores on the Learning to Teach Mathematics instrument placed her in the top 10% of all students participating in the study. Jayne saw herself as having the authority to determine the correctness of a mathematical answer and to make pedagogical decisions and wanted to help her students develop a similar sense of independence. Even as a preservice teacher, her view of textbooks provided a good example of her internal locus of authority: "I feel that [the textbook] gives a lot of good suggestions and activities for teachers to pick and choose from. Some activities just need a little adjusting." When asked how she planned her lessons, she noted, "I went through them [the curriculum materials] and saw what I thought. I went through and picked what I liked and what I didn't like and some stuff I thought was a good idea and just kind of modified it to what I thought my classes need."

As a classroom teacher Jayne challenged her principal's decision to ability-group students across classes for a portion of mathematics instruction every day to help students with weak procedural knowledge prepare for criterion-referenced testing. Jayne participated as required but continued to state her case. After test results came back and Jayne's students had the highest scores in the building, her principal relented and allowed her to manage mathematics instruction for her students for the full period.

A lesson typifying Jayne's teaching style and that of teachers fitting in the first cell of Figure 1 involved introducing first-graders to addition sentences. Jayne made the lesson engaging by using a song about frogs sitting on a log being joined by others. She allowed students to pick numbers and to act out problems, and then she introduced the corresponding notation. At one point a student proposed to act out a story that corresponded to 10 – 0. Jayne asked him to state his story as a number sentence, which he did correctly, and then asked him if it was an addition sentence. The child acknowledged that it was not. Jayne then asked him to recast his story so that it would correspond to an addition sentence, which the child did successfully. Throughout the lesson, Jayne asked questions such as: "What are we going to do with the numbers?" "How does the addition work?" "What can you tell me about adding two numbers?" "What is the number sentence? Why?" and "Does it make sense?"

In a post-observation conference following this lesson, Jayne reflected on her teaching, students' learning, and the content of the lesson. She articulated the purpose of her lesson this way:

> I wanted to see if the students could understand the concept of number sentences in a context, almost like a story problem. Because I have noticed that my kids have problems with story problems, you know, trying to decide if they need to add or subtract—knowing what to do with the story—where to start, I wanted them to have some basic comprehension of mathematical sentences and understand to go back and check their work, and to determine if their number sentence made sense—all the basics.

She also reflected on the seatwork portion of the lesson, noting that the individualized nature of the task allowed children to self-differentiate by selecting numbers with which they were comfortable. She noted that she challenged particular children by asking them to choose larger numbers.

Cynthia

Cynthia also possessed strong content knowledge in mathematics (and in other subjects, graduating from college with a perfect 4.0 grade point average). Cynthia considered herself to be "fairly good" at mathematics and attributed her success to teachers' clear explanations rather than her ability. She was often one of the first people to catch on to new material in her high school mathematics classes and her college mathematics class for elementary teachers, and her peers often came to her for help. If she did not understand something right away, she made a point of going to the teacher for help immediately.

However, Cynthia was a "teacher pleaser" with an external locus of authority both as a student and as a teacher. As a college student, Cynthia would often read ahead in the syllabus and ask detailed questions about expectations for future assignments. She frequently turned in drafts of assignments early and asked if she had done it "right." During interviews for this research study, Cynthia would ask, "Am I telling you what you want to know?" "Am I giving you what you need?" Cynthia's lessons, both as a preservice and an inservice teacher, were characterized by following the textbook and presenting clear, logical, well-organized explanations to students. Students engaged in activities, including hands-on activities, but these were always tightly structured and almost always procedural in nature.

Cynthia's interactions with students were generally quite directive. As a preservice teacher, she offered the following response to a written case: "To break this habit of writing the problems incorrectly, I would show them many times how to write the problem and have them do many practice problems themselves because practice makes perfect!" In providing suggestions for a peer who was having difficulty teaching a child the standard addition algorithm, Cynthia wrote, "First, the child needs to learn to start every addition problem in the right-hand column, or she will never remember to add the one. Model for her the correct way to do two-digit addition problems while you are teaching her how to actually do them." These examples typified her teaching during her field experiences and her first two years of teaching.

During one observation, Cynthia was circulating around the classroom as students worked independently. One student was confused, and she looked at his paper and said, "Write 3. Just write the number 3." Two more students had questions, and Cynthia told them what to write on the worksheet. Similarly, in a lesson on graphing in a second grade classroom, Cynthia was working with a small group of children who had just finished collecting data from their peers. She asked the group what color most people chose as their favorite. One student noted that he had a tie–the same number of people selected two colors as their favorite. Rather than asking the student what he meant by a "tie" or engaging the children in a discussion about what they should put on the worksheet if two colors were tied, she simply told the student to write the names of both colors in the blank. Later in the lesson, the worksheet contained the question "How many people liked horses or dogs the best?" The children were not sure how to interpret the "or" part of the question. Cynthia quickly told them to count how many people picked dogs and how many people picked horses and write the total in the blank.

Cynthia's reflections on her lessons were generally confined to assessments of student behavior and her organization in preparing for and implementing the lesson. For example, after one lesson she noted that students were engaged because "I made my word problems contain aspects of the Halloween season, such as trick-or-treating, candy, and toys, because those things were what [they were] interested in at the time."

Shelly

Shelly provides an example of a teacher with low content knowledge and an internal locus of authority. Shelly struggled with mathematics, taking the Praxis I basic skills test several times before passing the mathematics portion. She admitted to feeling a great deal of anxiety and to having a lack of self-confidence regarding mathematics teaching and learning.

Shelly demonstrated that she had an internal locus of authority by making modifications to textbook lessons even as a preservice teacher. During a field experience where she was in a school that had adopted a drill-oriented curriculum, she felt comfortable going outside the scripted lesson to add her own touches. In particular, she often incorporated children's literature and hands-on activities in her lessons. She also demonstrated her internal locus of authority in her reflections on her teaching. At the conclusion of an eight-week one-on-one teaching experience she noted about a student, "I have made some improvement through trial and error.... I stopped worrying so much about planning things that were on her grade level but rather planning things based on her success during a lesson."

Although Shelly planned lessons that engaged students, her lack of mathematical knowledge often caused her lessons to collapse. An illustrative example of how Shelly's weak mathematical knowledge interfered with her lesson planning came from a lesson on even and odd numbers that she taught in her first-grade classroom during her second year of teaching. After an introduction, Shelly handed the children pieces of paper and told them to write their first names on the paper. She then placed a pile of cubes at each table and told the children to count out enough cubes to equal the number of letters in their first names and snap them together. Shelly then told the children to "see if you can divide your cubes into two groups evenly." She chose a student's cube train and said "This is what I'd do." She modeled snapping off one cube at a time and alternating placing them in one pile or another. She then asked if the result was "fair," if the piles were the same (her informal definition of "even").

After giving the children time to work independently, she called the class back together and asked one child at a time to say how many letters were in his name, to show his cubes and tell whether he could put them into two equal groups, to declare whether his name (not the number of letters in his name) was odd or even, and to come to the front of the room to attach his name to a poster under "odd" or "even." She had one child snap his cubes off one at a time and place them in her hands, alternating between left and right. Then she had him count six in one hand and five in the other and asked him whether it was fair or even. The child said "yes," so she ended up telling him it was odd when he said the amounts were not equal.

The researcher observed that the students' textbook taught odd/even by having children use cubes to represent a quantity and then snap the cubes off in twos (pairs) rather than putting them in two groups (measurement division rather than partitive). When asked if she had already taught even and odd numbers this way, Shelly seemed completely unaware of this approach and said that she would try it because it might be easier. This is an illustration of Shelly feeling comfortable to go beyond the textbook to design her own lesson but not having the mathematical foresight to realize that she was creating problems with students' understanding by introducing a method that was contradictory to the method the students had to use to do their homework. While there is nothing wrong with introducing an alternative conception of even and odd numbers, what is problematic in this instance is that Shelly did not realize she was using a different definition and therefore had not thought through the consequences of the alternative definitions. In the post-observation conference Shelly indicated that she planned to delay the test on this chapter because she did not think the students had a firm grasp on even and odd numbers.

Tracey

Tracey was not confident in her mathematical ability, claiming that "math has always been my least favorite subject throughout school, and I've always called it my worst subject." She said she was not good at quick computations and memorizing algorithms, and this had worked to her disadvantage in school. She hoped not to have to take any math in college but said she enjoyed her math for elementary teachers course "once I figured out what I was supposed to be doing."

It appeared that Tracey had an external locus of authority for herself as a future teacher and that she saw herself as an external authority for her students. As a preservice teacher she noted that time would be a constraining factor in the classroom and that it would not be possible to hear ideas from many children. She stated that moving through the curriculum at a predetermined pace and adhering to school system requirements was a higher priority than listening to children. In other words, she was going to be the authority who would declare whether answers or approaches were correct or not. She also noted that sometimes what a child says is "off the wall and would only confuse matters more or lead the topic off on a tangent."

Observation of Tracey's first field experience showed that she did indeed view herself as an authority for her students. Her lesson was very directive, and she spoon-fed the children to enhance their short-term success. In helping second-graders complete a two-digit subtraction problem, she asked questions such as "Where do I start?" "What problem do I do first?" "What is eight minus five?" "Where do I put it?" "Am I done with this part?" and "Where do I go next?" Her lesson was a string of such bite-sized questions with no "Why?" questions at all. For much of the lesson she helped students create a bar graph about circus animals in order to answer word problems. She directed students: "Put your finger on the yellow box. Find the line that says 'elephants.' Point to the number of elephants. Who can tell me how many elephants there are?" Then she told a student to come to the board and put the prescribed number of elements on the graph she had created.

Tracey's overriding concern as a teacher seemed to be to "protect" children from the pain of doing mathematics by either breaking it down into tiny steps and asking structured questions that they could answer without much thought or by doing "fun" activities with them. An example of a fun activity occurred during a lesson when she was teaching kindergarten. The goal of the lesson was for children to tell time to the hour. She began the lesson by reading a children's literature book (*The Grouchy Ladybug* by Eric Carle) and then had the children make their own clocks in the shape of ladybugs using paper plates, construction paper, brads, and pipe cleaners. The bulk of the one-hour lesson was spent on the craft activity of constructing the ladybugs.

After they had spent 40 or more minutes on their creations, Tracey directed the children to fill in the numbers on the clock face. She did not anticipate that this would be difficult for kindergarteners (cognitively and in terms of fine motor skills), but most of the students ended up with inaccurate representations of clock faces (starting with 1 where the 12 should be, numbers unevenly spaced, numbers going past 12, etc.). Her plan had been to have the children bring their clocks back to the reading rug while she read the book again and have them show each hour on their own clocks. As the time allotted to the lesson began to wind down, she realized that the clocks were not functional because the clock faces were not accurate and because the crafts contained too much wet glue to withstand handling, so she omitted the final part of the lesson.

As she reflected later in the day, Tracey was disappointed in the lesson, but she mainly focused on the effort she had put into cutting out the ladybug wings and spots and bending construction paper to make ladybug legs along with her frustration that the clocks did not turn out to be functional because of the glue issue. She did not mention the fact that the clock faces were not accurate or that the lesson involved virtually no mathematics.

A similar but slightly less dramatic example comes from Tracey's second year of teaching second grade in a lesson on "greater than" and "less than." In an effort to make a lesson "fun," she gave each student three colored index cards—a pink one with a "less than" sign on it, a green one with a "greater than" sign on it, and a blue one with an "equal" sign on it. She put two numbers on the white board and asked the students to hold up the correct card (with the sign facing her). In planning this lesson she failed to take into account that the greater than and less than signs are really only one sign and that when the students held up the cards to face her, they would be facing the opposite of the way the children saw them at their desks. After about ten minutes, she called a halt to the lesson because she could not tell whether the children understood greater than and less than. In reflecting on the lesson, she noted that she had tried to make the lesson more fun and interactive than a worksheet but that a worksheet was probably a better way to assess this content.

Discussion

Although this manuscript presents examples of four teachers who seem to fit neatly into the cells of Figure 1, the reality is that teachers fit in one cell for some lessons and in another cell for other lessons. Teachers' practices are not consistent across time and setting, and the changes are not necessarily reflective of "growth" toward the upper left cell of Figure 1. For example, some teachers were very comfortable asking open-ended questions and building instruction from children's mathematical thinking during their first mathematics methods class field experience in which they worked with one child. In subsequent whole-class field experiences, however, they became more tied to the textbook and focused on correct answers. Although not explicitly documented in the cases reported here, it is also plausible that a teacher's content knowledge changes with respect to different topics, which may also influence the locus of authority or instructional practice to change.

While there is an abundance of research being done on various types of teachers' content knowledge, minimal attention has been paid to the notion of authority and ways to help teachers develop an internal locus of authority. The findings from this study suggest that there may be value in experimenting with and studying teacher education activities that attempt to shift locus of authority. Furthermore, this study highlights the potential value in studying the interplay between two simultaneous influences on teachers' practice.

The goal of proposing this framework is not to put teachers in boxes; rather, the goal is to call out, highlight, and draw attention to the interplay between content knowledge and locus of authority in order to illuminate aspects of practice that might be affected by these elements. The framework is a necessary oversimplification of complex reality in order to make a point and to help others identify prototypical cases. These prototypical cases can then help teacher educators consider how teacher education programs, induction year support programs, and professional development programs can leverage content knowledge and locus of authority to best assist teachers in developing a practice that leads to rich mathematical activity in the classroom.

References

Ambrose, R., Philipp, R., Chauvot, J., & Clement, L. (2003). A web-based survey to assess prospective elementary school teachers' beliefs about mathematics and mathematics learning: An alternative to Likert scales. In N. A. Pateman, B. J. Dougherty, & J. T. Zilliox (Eds.), *Proceedings of the 2003 Joint Meeting of PME and PMENA* (2, 33–39). Honolulu: CRDG, College of Education, University of Hawaii.

Bogdan, R. C., & Biklen, S. K. (1992). *Qualitative research for education: An introduction to theory and methods.* Boston: Allyn and Bacon.

Carle, E. (1977). *The grouchy ladybug*. New York: Harper Collins.

Eisenhart, M. A. (1988). The ethnographic research tradition and mathematics education research. *Journal for Research in Mathematics Education, 19*, 99–114.

Hill, H. C., & Ball, D. L. (2004). Learning mathematics for teaching: Results from California's mathematics professional development institutes. *Journal for Research in Mathematics Education, 35*, 330–351.

Kilpatrick, J. (1988). Editorial. *Journal for Research in Mathematics Education, 19*, 98.

Mewborn, D. S. (1999). Reflective thinking in preservice elementary mathematics teachers. *Journal for Research in Mathematics Education, 30*, 316–341

Zeichner, K. M., & Gore, J. M. (1990). Teacher socialization. In W. R. Houston (Ed.), *Handbook of research on teacher education* (pp. 329–348). New York: Macmillan.

Endnotes

[1] This study was funded by the Spencer Foundation under grant number 200000266. I am grateful to Patricia Johnson, David W. Stinson, and Lu Pien Cheng for their contributions to data collection and analysis throughout the project. An earlier version of this paper was presented at the 2009 annual meeting of the North American Chapter of the International Group for the Psychology of Mathematics Education and published in its proceedings.

Denise A. Spangler is professor of mathematics education and head of the Department of Mathematics and Science Education at the University of Georgia. Her teaching and research focus on preservice elementary mathematics teacher education and beliefs development. She can be reached at dspangle@uga.edu

Miriti, L., and Mohr-Schroeder, M. J.
AMTE Monograph 7
Mathematics Teaching: Putting Research into Practice at All Levels

5

Using Online Social Networking to Connect University Supervisors and Student Teachers

Landrea Miriti
Bluegrass Community and Technical College

Margaret J. Mohr-Schroeder
University of Kentucky

Student teaching is filled with day-to-day experiences that only cooperating teachers share with student teachers. The minimal interaction between student teachers and university supervisors is often a barrier to effective supervision. Online social networking can be a venue for connecting university supervisors to student teachers' daily triumphs and challenges. Engaging student teachers and university supervisors in online social networking presents a unique opportunity for the supervisors to monitor and support the development of beliefs and practices of prospective teachers throughout their student teaching experiences.

Student teaching is the capstone experience, and often the highlight, of programs for preservice secondary mathematics teachers. University supervisors, along with cooperating teachers, are charged with facilitating preservice teachers' learning during the student teaching experience. Student teaching is an opportunity for the supervisors to use "situational" teaching (Cohn, 1981) to help preservice mathematics teachers bridge theory and practice and implement reform-oriented, research-based mathematics instruction (Borko & Mayfield, 1995; Feiman-Nemser & Buchman, 1987). Yet research examining the supervision of student teachers reveals that supervisors often have minimal impact on student teachers' development, and may play a limited role in guiding their practices (Borko & Mayfield, 1995; Bush, 1986; Richardson-Koehler, 1988; Slick, 1998).

In a typical student teaching placement, university supervisors visit student teachers intermittently throughout the semester to make observations and provide feedback. They do not share common day-to-day classroom experiences with student teachers, as do cooperating teachers; thus university supervisors (hereafter referred to as supervisors) are often viewed as outsiders in the student teaching triad. Sporadic interaction between student teachers and their supervisors can be a barrier to effective supervision (Borko & Mayfield, 1995; Feiman-Nemser & Buchman, 1987; Richardson-Koehler, 1995).

To overcome this barrier, researchers have explored the use of new technologies to enhance communication between student teachers and supervisors (Holder & Carter, 1997; Pena & Almaguer, 2007). Online social networking can be a venue for connecting supervisors to student teachers' daily work and can provide unique opportunities for supervisors to monitor and support the development of prospective teachers' beliefs and practices throughout the student teaching experience.

An Online Social Networking Experiment

Secondary mathematics preservice teachers enrolled in a Master's with Initial Certification (MIC) program at a large university in Kentucky were required to keep an online journal of experiences during the semester-long student teaching experience. Preservice teachers used a blog tool available through Ning, an online social networking site (www.ning.com), which was free to users when the research was conducted [but converted to a fee based site in July 2010], to share reflections on classroom experiences. Student teachers were asked to reflect three to four times a week on classroom experiences and post reflections on personal Ning blog "pages" (see Figure 1). There were no assigned discussion topics or minimum requirements beyond posting three to four times per week. During the second half of the semester student teachers were asked to focus reflections on student mathematical thinking in the classrooms, although they were not limited to that topic.

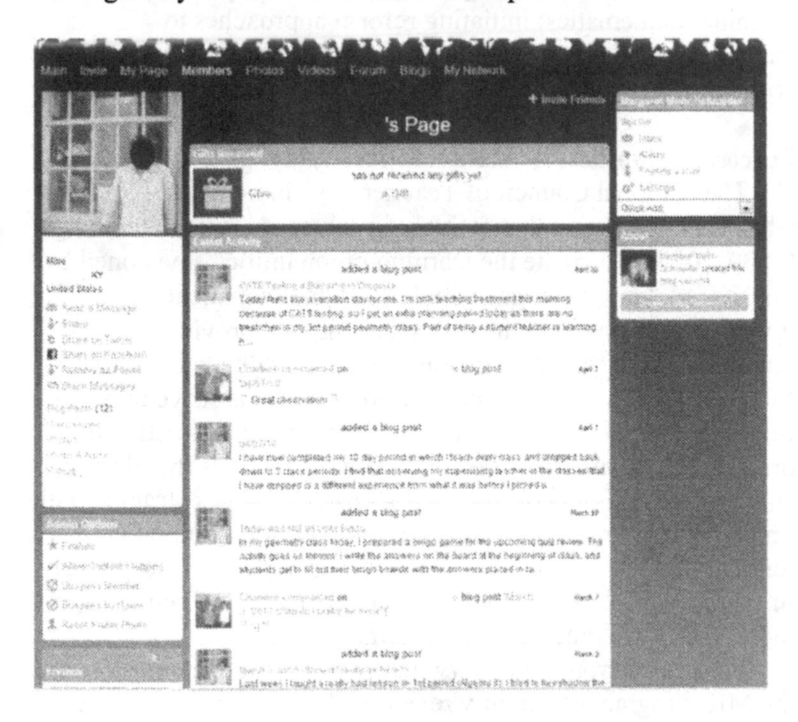

Figure 1: Screen shot of a secondary mathematics student teacher's page on Ning.

Access to the Ning site was limited to student teachers of mathematics, English, and science, methods instructors, and supervisors in order to secure a "safe" environment in case student teachers wanted to reflect on and negotiate conflicts, if any, with cooperating teachers. To ensure anonymity, students were not allowed to use student names or teacher names in their posts or discussions.

Reflections by Student Teachers

This section provides examples of student teacher reflections and supervisor responses regarding two issues specific to teaching mathematics: initiating reform approaches to mathematics pedagogy, and interpreting students' mathematical errors and misconceptions.

Perceived Conflicts in Mathematics Pedagogy

The National Council of Teachers of Mathematics [NCTM] advocates instruction that actively involves students in learning (2000, 2007). To create the learning communities envisioned by the NCTM, teachers engage students in mathematical investigations, facilitate inquiry learning, and provide opportunities for students to communicate mathematical ideas. Most mathematics teacher preparation programs strive to prepare future mathematics teachers to implement such instructional practices in meaningful ways. Yet research reveals that during student teaching, prospective mathematics teachers struggle with implementing this type of instruction and often abandon research- and reform-based approaches to instruction, especially in cases where their cooperating teachers model and encourage more traditional practices (Frykholm, 1996).

Blog posts from secondary mathematics student teachers in the MIC program commonly revealed discrepancies between theory learned in coursework and practices in the field.

- It does bother me a little to just lecture/give notes and do examples. I feel like I'm being a bad/boring teacher and going against everything we've been taught. However, I struggle in my mind because my [cooperating teacher] is a really good teacher, and her students learn the material, and they are advanced. (After all, that's how my advanced classes were, and I turned out ok.) But she does use more of a traditional style. (Posted by student teacher on January 22, 2009)

- I did talk to my [cooperating teacher] about doing activities in advanced geometry. She said what I thought was the case—you have to cover so much material, there isn't much time for investigations. (Posted by student teacher on January 23, 2009)

Several of the secondary mathematics student teachers felt obligated to replicate their cooperating teachers' instructional practices and subsequently felt conflicted when those practices contradicted the reform approaches introduced in the teacher education program. The social networking site provided an outlet for student teachers to air frustrations confidentially, and, in turn, provided a venue for supervisors to encourage student teachers' efforts to introduce research-based reform instruction. For example, the blog posts below document an exchange between a student teacher named Karey [pseudonyms used throughout] and her supervisor. Over the course of eight days, Karey moved from suggesting and experimenting with, to eventually implementing, meaningful research-based strategies to actively engage students in doing mathematics.

Karey: Today I had to give a practice ACT (only 23 questions) and then go through every single one of them with the students. I tried suggesting letting the students put the problems up on the board instead of going over each one myself, but my teacher did not want me to do that. I got the document camera out and worked through them one by one. I asked the students clarifying questions as we went along, but it was so boring. The poor kid right infront [*sic*] of me was falling asleep. I wanted to fall asleep. It was aweful [*sic*]. I told him he could get up and walk around for a second. He needed to. I felt bad for all of them. It was the 3rd time they had seen these problems and the 3rd time I had presented them. I hope that I never do this to my students, but today I did not have a choice. (Posted March 9, 2009)

Supervisor: In the future, you will have a choice. Hang in there! (Posted March 9, 2009)

Karey: With my Algebra 1 class we had a snowball fight today. On a half sheet of paper students had to write a difference of two squares problem. Next, students crumbled them up, through [*sic*] them around the room for thirty seconds, then had to open one and factor the difference of two squares. Then they crumpled up the paper, and we did it all over again. I thought my teacher was going to have a heart attack. It was amazing! The students actually did fairly well, and it was a great activity right before lunch.... Later on that day my teacher told me that she wants me to hold off on the activities in Algebra 1 for a few days. She said they need routine and structure and need to always know what to expect. I agreed to her request and will do as she asked, but I am really disappointed. These students never get the chance to do anything fun because she has never trained them in [*sic*] how to act appropriately when doing fun activities. Its [*sic*] sad. Oh well ... back to worksheets for a few days ... makes planning easy. (Posted March 11, 2009)

Supervisor: How brave of you to take the risk and try something different! I can understand your [cooperating teacher's] perspective but I think you make a very insightful observation … when you say that "these students never get a chance to do anything fun because [they] have never been trained in how to act appropriately when doing fun activities." … I agree that students can handle the wild fun things when given firm boundaries and ground rules. (Posted March 12, 2009)

Karey: So last night I knew that today I would need to do a review of trig identities before their quiz on Thursday, but I get so tired of doing example after example after example.... What I did today was split the class into 9 groups of three. Then we put the desks into 9 different stations. At each station there was a different type of review for thier [*sic*] quiz.... [One] was "Match Maker" and the students had to solve a trig equation and match the problem to the solution. Another station was called "Two Truths and a Lie" where the students had to verify three trig identities, but one was false and would not work out. The list goes on and there was a cute name for each:) They stayed at each station for three and a half minutes and then would have thirty seconds [to switch]. It was great to see all of the students engaged and doing their thing. I hated myself last night at 11:30 pm when I was still working on it, but at 9 am this morning I loved it because they were doing all of the work, and I was just walking around making sure everything was running smoothly. Yay cooperative learning! (Posted March 17, 2009)

Supervisor: Bravo, Karey for thinking creatively, breaking out of the routine and using stations as a way to review! You go girl!... I know that preparing stations was a lot of work but the pay off is worth it. (Posted March 19, 2009)

It is evident Karey was committed to finding ways to engage her students. Her posted reflections range from expressions of frustration with initial attempts to descriptions of radical efforts (e.g., snowball fights) to change the climate in her classroom. The supervisor supported Karey by empathizing with her frustrations, affirming her efforts, and applauding her successes. The online exchanges between Karey and her supervisor helped to sustain Karey's efforts to incorporate strategies to engage students actively.

The crucial supporting role played by Karey's supervisor via online social networking reflects the findings of Gwyn-Paquette and Tochon's (2002) research on how preservice second language student teachers navigated through the difficulties of introducing cooperative learning in classrooms. Although student teachers' "convictions about the usefulness of the cooperative approach and other personal motivation provided the springboard for experimentation, it became evident that … expert coaching and continuous moral support are essential to foster the development of pre-service teachers' ability to innovate their teaching approach" (p. 204).

Negotiating Innovative Strategies

Prospective mathematics teachers in the MIC program were encouraged by their university methods instructor to engage students in mathematical investigations during student teaching experiences. Lappan (1997) describes investigations as "posing problems and questions, and then skillfully guiding problem solving and discourse so that students' ideas are constantly probed and pushed towards more powerful mathematical realizations" (p. 210). Effectively navigating a classroom of students through a mathematical investigation is a daunting task and is especially challenging for novice teachers. The Ning site provided a window through which supervisors could monitor student teachers' attempts to incorporate investigations in their student teaching practice. In the blog post below, mathematics student teacher Robin describes her first attempt at teaching through investigations:

> I wrote my own investigation. I was excited to incorporate group work and then I wanted students to present their findings—spurring mathematical discussion with/to their

> peers.... But it did not work out as I had hoped. In fact, it made me not want to do investigations... I felt they would have understood better if I just told them the theorem and explained by examples.... I haven't given up on discovery learning, but it sure does take longer and can be more confusing. (Posted by Robin on February 18, 2009)

The social networking site was a non-evaluative environment where Robin could share with peers and supervisors her disappointment and doubt about teaching with investigations. Later in the same blog post, Robin shared the significant lessons she learned as a result of her failed first attempt at implementing an investigation.

> *Lesson learned*: — Choose my investigations wisely and make sure it [*sic*] will contribute to overall understanding. — Need time to debrief or investigation may be worthless. — Students need to be trained to work in groups. — This made me want to lead investigations from the front of the class. — Unless, the investigation is easy or you have a smart class, you need to focus on transitions and give an intro to the investigation. (Posted by Robin on February 18, 2009)

The supervisor's response affirmed Robin's "lesson learned" and challenged her to reflect on and revise her practice.

> The great lessons learned will definitely help you when designing the next investigation. Glad to know that you are not giving up completely on doing investigations. Challenge yourself to figure out how to make it work better. (Posted by supervisor on February 18, 2009)

The following message, posted nine days later, indicates Robin had begun to realize the potential of teaching with investigations. She reported on her success facilitating a geometry investigation activity. More importantly, her reflection revealed she had developed a practical and yet sophisticated understanding of what investigation and discovery learning could look like in a mathematics classroom.

> I had an investigation that Ms. B uses with her regular Geom [*sic*]. When we first started it, I was afraid it was too elementary for advanced geom. [*sic*], but it went really well…. Hopefully, the investigation will help them remember the relationships better. I like to work small discovery activities into almost every lesson. Normally when I think of investigation, I think of something long and drawn out, working in groups, but it doesn't have to be like that. They can just be short like this one. Also, I realized that a lot of these "discovery" activities really just show where things orginated [*sic*] and let students discover relationships building on their prior knolwedge [*sic*], so I realized that that is what my [cooperating teacher] does when she explains things. She doesn't have a formal activity where they measure sides of a triangle, for example, but the way she explains it is like a discovery activity. (Posted by Robin on February 27, 2009)

As an active participant in the online dialogue, the supervisor affirmed and summarized the student teacher's construction of knowledge about investigations and discovery learning.

> You really picked up on a key idea by realizing that investigations can be short and effective and that sometimes just the way a teacher explains things can be like discovery learning when the explanations are not just about telling but helping students to put the concept together. (Posted by supervisor February 27, 2009)

Robin's evolution from doubt about having students derive concepts through investigations to expressing desire to "work small discovery activities into almost every lesson" is an example of pre-service teacher growth that surfaced for the supervisor's viewing and feedback through online social networking.

Examining Mathematical Content in Teaching

In addition to providing opportunities to monitor and support student teachers' developing mathematical pedagogy, online social networking was a catalyst for examining mathematical content for teaching. The student teachers' blog posts often described encounters with student misconceptions of mathematics concepts. For example, Dana voiced her frustration with her students' incorrect application of the "cancelling" procedure. In response, Dana's supervisor shared her own approach to helping students understand the underlying mathematical concepts that justify the cancelling procedure.

Dana: It wasn't until we started doing examples that I realized the problem with cancelling. I have never liked the term because it doesn't have any logic behind it. Students do not know why they are cancelling. Because of this, students started cancelling common terms instead of common factors. I tried to explain why it was not right and we even plugged numbers in to see why it did not work. Most students were able to see the reasons, but I noticed on homework others did not. I am not sure what else to do. These students have heard the word cancelling for a long time and can't seem to shake the idea. (Posted April 6, 2009)

Supervisor: Every teacher has encountered just what you described about students cancelling common terms instead of common factors.... I try to overcome this misconception by telling students that we do not cancel but when we have cases where some form of 1 [is] ... multiplied by "something" (because 1 times anything is the thing (multiplicative identity property)), we can just write the "something". So the words I use in class are not, "can we cancel" but instead, "do you see a form of 1 and is it being multiplied by something?" It [sic] think these words really get at the concepts behind the procedure of crossing out the same thing in numerator and denominator. (Posted April 7, 2009)

In a subsequent post, Dana described her experience with tutoring a student after school describing the student's thinking as "mind boggling." In response, the supervisor encouraged Dana to focus on the "whys" behind the student's errors.

Dana: We were trying to simplify rational expressions. After leading the student through all the steps in factoring we got to a point where she had to simplify 3p/3p.... She said that the answer was 11. WHAT???? HOW DID YOU GET 11!! She explained that 3/3 was 1 and p/p was 1 and 1 and 1 was 11. I almost jumped out of the third story window. (Posted by Dana on April 9, 2009)

Supervisor: Wow, this thinking is so interesting and actually very understandable! Doesn't a 1 next to a 1 mean 11? Makes sense to me. Before you step out on the window ledge, ask your self [sic] what is missing in this student's understanding that made her say 11. Perhaps what is the missing [sic] is the understanding that the operation between 3/3 and p/p is multiplication which is not trivial idea [sic] because there is so much going on, so much to interpret, in the expression 3p/3p. (3*p divided by 3*p, or 3/3 times p/p) (Posted April 9, 2009)

The above examples illustrate that online social networking between MIC secondary mathematics student teachers and supervisors presented opportunities, grounded in student teachers' interactions with real students, to attend to the nature of mathematics and how students learn mathematics.

Conclusions and Implications

To "shape" what prospective teachers learn from student teaching, teacher educators and supervisors must be "actively present in student teaching to provide prospective teachers a concrete sense of pedagogical thinking and acting" (Feiman-Nemser & Buchman, 1987, p. 272). As illustrated in this paper, online social networking enables supervisors to be actively present in student teaching to support secondary mathematics student teachers' particular efforts to introduce reform mathematics pedagogy, to facilitate secondary student teachers' understanding of students' mathematical thinking, and to address misconceptions. Online social networking provides an opportunity for supervisors to move beyond the responsibilities of occasional observer and assessor to the role of ongoing coach and mentor.

The 24-hour access to communication provided through online social networking gives supervisors an opportunity to share and respond to student teachers' daily triumphs and challenges. Moreover, online social networking provides a continuously documented conversational venue for supervisors to assess, affirm and challenge prospective secondary mathematics teachers' beliefs and instructional practices as those beliefs and practices emerge from daily student teaching experiences. Follow-up interviews with student teachers about use of the social networking environment and views of relationships with supervisors should be conducted in order to get an in-depth picture of the effectiveness of limited access social networking environments. Future research should explore the role of online social networking in helping supervisors monitor the development of student teachers' beliefs and practices.

References

Borko, H., & Mayfield, V. (1995). The roles of the cooperating teacher and university supervisor in learning to teach. *Teaching and Teacher Education, 11*, 501–518.

Bush, W. (1986). Preservice teachers' sources of decisions in teaching secondary mathematics. *Journal for Research in Mathematics Education, 17*, 21–30.

Cohn, M. (1981). A new supervision model for linking theory to practice. *Journal of Teacher Education, 32*, 26–30.

Feiman-Nemser, S., & Buchmann, M. (1987). When is student teaching teacher education? *Teaching and Teacher Education, 3*, 255–273.

Frykholm, J. (1996). Pre-service teachers in mathematics: Struggling with the standards. *Teaching and Teacher Education, 12*, 665–681.

Gwyn-Paquette, C., & Tochon, F. V. (2002). The role of reflective conversations and feedback in helping preservice teachers learn to use cooperative activities in their second language classrooms. *The Modern Language Journal, 86*, 204–226.

Holder, J., & Carter, D. (1997, December). *The role of new information technologies in facilitating professional reflective practice across the supervisory triad.* Paper presented at Annual Conference of the gasat-IOSTE, Perth, Western Australia.

Lappan, G. (1997). The challenges of implementation: Supporting teachers. *American Journal of Education, 106*, 207–239.

National Council of Teachers of Mathematics. (2000). *Principles and standards for school mathematics*. Reston, VA: Author.

National Council of Teachers of Mathematics. (2007). *Mathematics teaching today.* Reston, VA: Author.

Pena, C., & Almaguer, I. (2007). Asking the right questions: Online mentoring of student teachers. *International Journal of Instructional Media, 34*, 105–113.

Richardson-Koehler, V. (1988). Barriers to effective supervision of student teaching: A field study. *Journal of Teacher Education, 39*, 28–34.

Slick, S. (1998). The university supervisor: A disenfranchised outsider. *Teaching and Teacher Education, 14*, 821–834.

Landrea Miriti is an associate professor of mathematics at Bluegrass Community and Technical College in Lexington, KY. She is a member Cohort 3 of the NSF-funded CLT, ACCLAIM and is a Ph.D candidate at the University of Louisville. She can be reached at landrea.miriti@kctcs.edu

Margaret Mohr-Schroeder is an assistant professor of mathematics education at the University of Kentucky where she is chair of the Secondary Mathematics Program. She can be reached at m.mohr@uky.edu

Cwikla, Julie
AMTE Monograph 7
Mathematics Teaching: Putting Research into Practice at All Levels
© 2010, pp. 73–90

6

Using Collegiate Classroom Video: Mathematics Faculty Reflect on Their Own and Their Peers' Practices

Julie Cwikla
University of Southern Mississippi Gulf Coast

Mathematics education faculty members from five institutions of higher learning participated in a professional community of practice led by the author. Members' classroom practices were shared and critiqued by the community via video clips and analysis. The video analysis indicates that (a) classroom video can be an effective medium for professional discussion; (b) faculty members in higher education are a diverse group of professionals in their views of teaching and learning, practices, and reflection; (c) faculty need to be exposed and educated about other methods of teaching and learning; and (d) faculty members' views of teaching and learning, their actual classroom practice, and their reactions and comments about peers' classrooms are highly aligned. Observing, recording, and analyzing practice as a community is an intimidating task, but the prospects appear promising.

The National Council of Teachers of Mathematics (NCTM) *Principles and Standards for School Mathematics* (2000) and *Mathematics Teaching Today: Improving Practice, Improving Student Learning* (2007) support learning environments that are contextual and meaningful for learning, student-centered, and participative in structure, involving the learners in their education. At the postsecondary level, King (1992) reported that long-term retention and understanding of mathematics content for undergraduate students are most likely to develop from student-centered question-generating exercises in the collegiate classroom. An interview study of university students also supported student-faculty interactions, concluding that for students' learning, meaningful classroom interactions with the professor ranked as one of the six most important classroom features (Clarke, 1995).

However, in a study of university faculty and the stresses in their profession, interaction with students brought about the most angst in the profession (Gmelch, Wilke, & Lovrich, 1996). The stress of student interactions combined with varying faculty views of learning might contribute to dissonance between students' and faculty members' priorities and goals for their relationships and interactions. Similar to findings for K–12 teachers (Cwikla, 2002), university faculty members' views of teaching and learning range from teaching as transmission to student-centered construction of knowledge (Cwikla, 2008; Samuelowicz & Bain, 1992).

Mathematics educators can develop a better rapport with their students and better serve them as learners and future teachers. How mathematics educators learn about effective practices in a situated manner is the focus of this chapter. The larger body of work from which this study is drawn focuses on a "holistic rather than atomistic" view of professional learning (Webster-Wright, 2009, p. 728).

Paulsen and Feldman (1995) suggested that faculty naturally develop as reflective practitioners through feedback from (a) students, (b) colleagues, and (c) consultants if they are embedded in a culture supportive of teaching improvements. And Kranier (2001) stressed the importance of a community consisting of a "network of critical friends" (p. 289). As alluded to by Kranier, a *community of practice* can be used to characterize a professional group of learners with a common goal such as educating young people; a set of norms, expectations, and standards; and a method or manner to systematically share information about their practice (Wenger, 1998).

Developing a community of practice is a process that requires a long-term vision, a shared mission and desire to improve, and commitment by the participants to the goal of questioning the status quo and finding ways to improve for the sake of students' learning. These features of faculty development guided the development and structure of the Professional Mathematics Educators' Forum (referred to as a community of practice hereafter). The mission of the community of practice in this study is to use data collected in collegiate classrooms to learn from and critique each other for the purposes of developing an understanding *of* practice, and applying that understanding *in* practice. Understanding of practice must be integrated with teachers' understanding in practice (Ball & Cohen, 1999; Dall'Alba & Sandberg, 2006).

Context

Greenwood University (pseudonym) is a four-year institution with a student enrollment of about 14,000 undergraduate students, many of whom transferred from community and junior colleges in the surrounding area. Over 70% of the future teachers who graduate from Greenwood have had at least two semesters of mathematics at four feeder institutions. The 18 mathematics faculty members who participated in the community of practice were from these five institutions. There were 16 participants from mathematics departments, and all but two held master's degrees in mathematics, computer science, or statistics. The other two participants were doctoral-level mathematics educators. The majority of participants had experience teaching in the public school system.

Before Greenwood University students pursuing a degree in elementary K–8 certification take a mathematics methods class, they must complete four mathematics courses: College Algebra, and Mathematics for Elementary Teachers I, II, and III. This study investigated the practices of nine faculty members in the community of practice who teach College Algebra, which represents the first mathematical contact with preservice elementary teachers.

Method

Prior to development of the community of practice, mathematics faculty from the five institutions met sporadically to discuss course alignment and articulation, but rarely as a whole group and with few clearly defined goals. The community of practice provided professional opportunities for all to meet, share best practices, discuss course activities and specifics, demonstrate use of various technologies, and share classroom video data. Members were compensated for mileage, time spent in professional meetings, and survey completion.

Mathematics Faculty Survey

Prior to the first community of practice meeting, the 18 participating faculty members completed a survey that addressed their own teaching experience and educational background and their views of learning and teaching. They were also asked to select from adjectives describing their preservice teachers. The views of learning and teaching portion of the survey consisted of 18 items constructed to classify faculty members' views into two entities: (1) student-centered, and (2) curriculum-centered. A factor analysis of the 18 items was conducted and a detailed analysis can be found in Cwikla (2002). Two items from the views of learning and teaching portion follow: (1) Learners benefit more if they solve a problem on their own than if they follow someone else's method; (2) It is better to plan learning opportunities that build on what I think students already understand than just letting things happen spontaneously. Responses for each item ranged from 1 (disagree) through 4 (agree). Total scores ranged from 18–72 points—the higher the score, the more student-centered the respondents' views. Faculty members' views of teaching and learning were used as a lens to investigate classroom practice and understand the connection among self-reported data, actual practice, and critique of peers.

Classroom Video Collection and Coding

All participating faculty and their classrooms were video recorded in the 2004 fall semester. Two digital video cameras were used. One camera was placed on a tripod at the back of the room and focused on the professor or instructor (here after referred to as "professor"). The second camera was used by a roaming camera operator and focused on the students, including their interactions, note taking, participation, and norms.

The coding scheme used to analyze the classroom video data was based on the work of Gearhart et al. (1999). The "Integrated Assessment" and "Conceptual Issues" codes were applied to each video by two independent coders. The Integrated Assessment code "captures student opportunities to participate in classroom discussions built upon mathematical thinking" (Gearhart et al., 1999, p. 297), and the Conceptual Issues code "captures student opportunities to engage with conceptual issues underlying problem solving" (Gearhart et al., 1999, p. 297).

Faculty reflection on their own classroom video. All videotaped faculty members were provided a copy of their professor-focused video and asked to complete fifteen reflection questions about what they learned by watching their own classroom. Faculty members were given the opportunity to watch themselves before clips of their lesson were shown to the entire group. Providing the video in advance to the nine faculty members gave them a chance to view their own work before it became public and served as a source of reflection data.

Public sharing and evaluation of classroom videos. The goal for the 2005 spring community of practice meeting was to watch and evaluate video clips from College Algebra classes taught by nine of the faculty members. For each class, a five- to seven-minute video clip typical of the lesson was selected. Each community of practice faculty member was given nine separate feedback sheets, one for each video. Six feedback questions guided reactions to each video clip. The name of the faculty member providing feedback was printed on the top of the sheet with the notion this would (a) tame harsh criticism, (b) make faculty members accountable for their feedback, (c) help to create as collegial an atmosphere as possible, and (d) provide contact information if the videotaped faculty member had questions about the comments. Although the non-anonymous format of the feedback might have distorted faculty members' responses, it helped develop and support the community of practice environment. "Feedback," "guidance," "suggestions," and "help" were terms used to clarify the formative purpose of the activity and encourage productive suggestions for improvement.

Results

Faculty Survey Data

The 18 faculty members' years of teaching ranged from six to 30. The views of learning and teaching mean score was approximately 50 points with a range of 25, from 38 to 63. There were no detected correlations between areas of degree specialty (mathematics education, mathematics, computer science, or statistics) and views of teaching and learning or between years of experience and views of teaching and learning. The most common adjectival descriptors for typical preservice teachers were: math-phobic, weak in math, risk taker, conscientious, and arithmetically incompetent.

Classroom Video Analyses

Each of the College Algebra student videos were analyzed by two reviewers and coded using a Likert scale from 1 (disagree) to 4 (agree) for each of "Conceptual Issues" and "Integrated Assessment." The nine classrooms fell roughly into three groups of three, which are classified by the following student roles.

Group 1: Note taking (G1). Three classrooms consisted of a traditional lecture format with limited student-teacher or student-student interactions. In these classes students generally looked from their notebooks to the board, copying what the teacher wrote in a rote fashion. The few questions asked by the teachers of Group 1 were generally leading, directed, and procedural requiring a yes/no response.

Group 2: Practice and assessment (G2). Three teachers combined traditional lecture format with some student seatwork and individual or paired activities as the teacher walked around the class answering questions and giving direction. In general, the teacher's assessment or questioning of the students was procedural but required significantly more student participation than in Group 1. Students knew that they would be verbally assessed, observed, and questioned on a daily basis in these courses as they interacted with the teachers. Student participation and active attention to the work were the norm.

Group 3: Prepare and problem solving (G3). Teachers in these classes typically began with what one referred to as a "mini-lecture" followed by a hands-on activity with manipulatives. Each method for teaching included student work in groups or pairs, encouraged and required student participation and explanations, and required students to present and share their findings with the class. The dialogue and assessment of students' thinking were not consistently conceptual, nor were students always required to understand more than their own methods.

Two of the three Group 3 classes were offered by faculty members at the four-year institution. These two faculty members also had the highest views of learning and were the most student-centered in their perceptions. There was consistency across self-reported views of learning and subsequent observed practice.

Individual Faculty Video Reflection

The nine college algebra faculty members were each given 15 reflection items to address after watching their student-focused video recording. One question was, "What surprised you about your practice?" The following types of answers are listed according to their level or depth of reflection on classroom practice, from least to greatest. The teacher's views of learning score and the video code are included. A Group 1 (G1) teacher reflecting on her video wrote, "I learned that they were very attentive and listened well" (Eve, 38, G1). Other responses included: "I did not learn this but I do know that in the ideal situation there should be more student participation" (Kate, 39, G1); "I did not realize how much I talk with my hands!" (Sandy, 47, G2); and "I am really quick to ask 'Are there any questions,' and quickly say 'OK' and progress with the lesson" (Olivia, 60,

G3). These last three comments indicate a progression, from Kate's expression of distance from her act of teaching, to Sandy's attention to herself with a relatively topical issue (her hands), to Olivia's thoughtful critique of her questioning methods. These answers demonstrated ways that the teachers reflect on their teaching related to their views of teaching as well as their practices.

Another question encouraged reflection from an outsider's point of view: "What might another mathematics educator learn from watching this video?" Teachers from all three groups shared the following: "I would hope that a new teacher would see how to be organized and present the problems in an orderly manner on the board" (Eve, 38, G1); "They may see things that I don't see that would help them be a better teacher. They could even give me some ideas of things I could do differently to be a better teacher" (Julia, 46, G2); and "I try to create in my classes an open atmosphere in which students feel comfortable asking questions … try to engage many different students in a class discussion by asking particular students probing questions" (Joanna, 60, G3).

The last reflection prompt was "What immediate changes/ adjustments/improvements might you make to your teaching practice in general as a result of watching this video?" Answers included: "I did not notice anything that I would improve" (Eve, 38, G1); "I always hope to do a better job next semester…. Any changes that I make are usually minor changes" (Kate, 39, G1); "I am going to begin incorporating problems that students must work together in small groups to solve at some point during my lecture" (Julia, 46, G2); and "I plan to spend more time prior to the class planning ways of providing students opportunities for small group interaction" (Joanna, 60, G3). All reflection questions were completed individually by the videotaped faculty members and were based only on the video of their own classroom.

Making Practice Public

Twelve faculty members who were present at the 2005 spring community of practice meeting studied each of the nine College Algebra video clips. The video clips were presented in alphabetical order and viewers were not aware of the earlier coding procedures. Faculty members were asked to respond to questions for each video:

1. Briefly describe the learning environment in this classroom.
2. What are these students learning and how do you know?
3. What have you learned from watching this video that you might try in your own class?
4. How and in what ways does this lesson challenge the students' thinking? Be specific.
5. As a peer what advice would you give this teacher about gauging her students' understanding of the lesson?
6. What could this teacher have done differently to further engage her students in the content/lesson?

A selection of the feedback comments from Items 1 and 2 describing student learning and classroom environment are found in Table 1. Similarly, the faculty members' suggestions for improvement captured in questions 5 and 6 are reported in Table 2. The comments are organized by group to illustrate how, for example, teachers in Group 1 reacted to Group 3 classrooms and vice versa.

Table 1

Summary of Responses to Feedback Items 1 and 2 Describing the Learning Environment

		Classrooms Being Observed		
		G3 – Prepare & Problem Solve	**G2 – Practice & Individual Assessment**	**G1 – Note Taking**
Teacher Groups Providing Feedback	**G3**	• Very cooperative. • Students doing assigned group work after problem is presented. • Group work with graphing calculator • Students are interacting and discussing the appropriate scale, etc. of the graph. Teacher walked around the room assisting individuals or small groups of students. • Individual attention – let students help one another.	• Teacher centered with student interactions • Students are actively engaged in open discussion concerning values selected for domain. • Answering questions, telling the teacher what the next step is. • Teacher presenting examples on overhead • Teacher centered with students' interactions. Students are using a graphing calculator. • After presenting example, gives students another example to interpret – Several students respond correctly to questions.	• Traditional • Students are taking notes and seem to be following examples. • Teacher using overhead. Working example of synthetic division on whiteboard. • Teacher lecturing, students copying examples, there is student/ teacher interaction. • Teacher centered with some students interactions.

Teacher Groups Providing Feedback		Classrooms Being Observed		
		G3 – Prepare & Problem Solve	**G2 – Practice & Individual Assessment**	**G1 – Note Taking**
	G2	• Student centered group work • Students working in groups. Not all groups had same problem. All had same type of problem. Students helping students. Students feel at ease • Students are talking to each other and helping each other realize how to change windows so they can see max and min	• Students are responding to questions • Teacher is lecturing with use of overhead projector. She's interacting with students through asking • Teacher Lecture – Then students working at desk. Teacher going around the room to observe students work.	• Lecture and demonstration • Students copied examples from the board. • I didn't hear many questions so I don't know how much they were actually learning. I think she was introducing this and they were just copying for now.
	G1	• Looked like lots of learning going on • Very informal, comfortable, not so teacher driven	• Traditional lecture, good interaction with students • Traditional lecture	• Very teacher oriented • Traditional lecture. Good quiet environment. • It was a pleasant and comfortable feeling classroom. • She was very prepared, but also tried to get her students involved.

Table 2

Summary of Responses to Feedback Items 5 and 6 Offering Suggestions for Improvement

		Classrooms Being Observed		
		G3 – Prepare & Problem Solve	**G2 – Practice & Individual Assessment**	**G1 – Note Taking**
Teacher Groups Providing Feedback	**G3**	• Students had ample opportunity to be engaged in lesson through individual class work and open discussion. • Students were totally engaged • After group work is completed, have some or all groups share results using overhead calculator.	• Continue using her questioning techniques – this allows her to provide immediate feedback Have students work in groups and interact with individual groups. • Maybe students could work in "pairs" then some could share for a change of pace • Ask them "Why?" • Ask your students to explain and tell you what to do next • Call on the students by name when asking questions – Then wait for them to answer.	• Talk more directly to them – not so much to the board. • Be sure to ask probing questions about the details of the problem presented. Relate algebraic steps to similar steps in an arithmetic problem. Make connections. • Hopefully students are "thinking" as they take notes but they could be rotely copying what is presented. • Ask students to share answers after question is presented • Ask your students questions. Let them tell you how to solve or have them come to the board and solve. • Have the students tell you what to do next. … Use manipulatives, relate to real world

		Classrooms Being Observed		
		G3 – Prepare & Problem Solve	**G2 – Practice & Individual Assessment**	**G1 – Note Taking**
Teacher Groups Providing Feedback	**G2**	• Wonderful job! You have inspired me to give group work a concentrated effort! • The students were very engaged. Even when she wasn't helping one, they were helping each other!	• She did a great job. She even noticed one student who was not at the right place and helped him get back on track.	• Allow more time for student questions. Make it a point to ask them leading questions if they have none. • Ask leading questions that force students to answer. This way the teacher will know if students are really learning. • Create opportunity for interaction
	G1	• *No comments were made by any of the G3 participants about the G1 classes and what they could do to improve.*	• She asked questions. She made them think about what points of a graph really meant. • I think a student here has a very positive experience	• It seemed to be a formula driven assignment. I heard "b over a", etc. • I think for a class of that size, the students were extremely involved in the lesson • Look at the students more – but this was a very short clip • Ask more questions of the class • She continued to involve her students – good job

Emergent Patterns and Implications

Tables 1 and 2 reveal that Group 3 offered more comments describing the learning environments and more suggestions for improvement than the other two groups. Also, there is consistency in the way the entire group describes the classrooms taught by instructors in Group 1. All the descriptors are in the range of "teacher-centered," "lecture," and "traditional." This suggests that the community of practice members share a definition of traditional lecture format.

In the left column of Table 1, the Group 3 lessons are described by the community of practice. The Group 2 and Group 3 educators use phrases such as "cooperative," "students interacting," and "individual attention" to describe these classrooms. By contrast, the Group 1 respondents' comments are generally vague and do not provide accurate descriptions. For example, the lecture-leaning teachers wrote, "looked like lots of learning going on" and "not so teacher driven." This may be because: (a) this was the first time these individuals had actually seen "group work"; (b) they did not have words to describe what they were seeing; or (c) they did not agree with or approve of the type of teaching they were seeing and therefore did not want to write much about it.

Peer Critique and Suggestions

The community of practice in general had more to critique and more advice to offer teachers in both Groups 1 and 2. This reflects a consensus that there is a need for improvement in both of these groups, or perhaps it was easier to make discrete and specific suggestions because the lecture format was more familiar. The most diverse and pointed suggestions for Group 1 came from Group 3. Their suggestions focused on interacting with the students, such as "talk more directly with them," "ask students to share answers," and "ask probing questions." These types of interactions were present in all of the Group 3 lessons. Similar comments were offered for Group 2 but were more specific, encouraging interactions among the students: "maybe the students could work in pairs" or "have the students work in groups." These comments suggested that Group 3 recognized that Group 2 teachers already interact with their students but need to foster more student participation. The Group 3 comments for other Group 3 members consisted mostly of compliments and encouragement as opposed to formal or specific suggestions.

Teachers in Group 2 also had only compliments for Group 3 lessons. One Group 2 educator, Becky, wrote "You have inspired me to give group work a concentrated effort." This illustrates anecdotally the power of video prompts for professional reflection and growth. Group 1 had no comments for Group 3, even though they watched three different clips, and there were three respondents. In other words there were nine opportunities for Group 1 to give suggestions to Group 3, and none were forthcoming.

Reflections and critiques of peers' practice provided a glimpse into the thoughts and reactions of educators regarding types of learning environments different from their own practice. Participants from Group 3 were more precise in language and specific in their suggestions about ways to improve. In addition, Group 3 participants clearly recognized differences among the three different practices. Some Group 2 educators displayed a desire to improve in their reflection comments. For some, this was the first time they viewed student-centered methods in a collegiate classroom, and they felt comfortable enough to express this both in writing and in the group discussion that followed. Group 1 educators, the most traditional in classroom practice and in their views of learning and teaching, were the least descriptive in their comments and indications that they would attempt to change their practice.

Discussion

Mathematics faculty members in this study were willing and enthused about their participation in the study. Some were apprehensive about the prospect of being video recorded in their classrooms, but in the end they were comfortable with the overall experience and sharing video clips with peers. The supportive climate and the culture of growth and development for all participants helped curb faculty members' apprehensions.

The video analysis by the community of practice indicates that (a) classroom video can be an effective medium for professional discussion and comparison in higher education; (b) mathematics faculty members in higher education are a diverse group of professionals in their views of teaching and learning, classroom practice, and levels of reflection; (c) faculty members in higher education need to be exposed to and educated about other methods of teaching and learning; and (d) faculty members' views of teaching and learning, their actual classroom practice, and their reactions and comments about peers' classrooms are highly aligned.

In conjunction with the results of this study and in potentially generalizing to a larger population, mathematics educators should consider the entire collegiate mathematics experience of preservice teachers. The first collegiate experience of future teachers may be a course like College Algebra. How this and other content courses impact these future teachers' visions of mathematics teaching and learning, as well as how these courses can be improved by supporting continued professional learning for faculty in higher education, are issues that deserve more attention. Observing practice is a daunting task, but the prospects appear promising.

References

Ball, D. L., & Cohen, D. K. (1999). Developing practice, developing practitioners: Toward a practice-based theory of professional education. In L. Darling-Hammond, & G. Sykes (Eds.), *Teaching as the learning profession: Handbook of policy and practice* (pp. 3–32). San Francisco: Jossey Bass.

Clarke, J. A. (1995). Tertiary students' perceptions of their learning environments: A new procedure and some outcomes. *Higher Education Research & Development, 14*, 1–12.

Cwikla, J. (2002). Mathematics teacher's report about the influence of various professional development activities. *The Professional Educator, 24*, 75–94.

Cwikla, J. (2008). Lifelong learning: Mathematics faculty work to improve their practice. In M. Qazi (Ed.), *Proceedings of the 5th Annual TEAM-Math Partnership Conference Pre-Session*. Tuskegee, AL: Tuskegee University. Retrieved from http://www.team-math.net/tuskegeeconference/proceedings/index.html

Dall'Alba, G., & Sandberg, J. (2006). Unveiling professional development: A critical review of stage models. *Review of Educational Research 76*, 383–412.

Gearhart, M., Saxe, G., Seltzer, M., Schlackman, J., Fall, R., Bennett, T., Rhine, T., & Sloan, T. F. (1999). Opportunities to learn fractions in elementary mathematics classrooms. *Journal for Research in Mathematics Education, 30*, 286–315.

Gmelch, W. H., Wilke, P. K., & Lovrich, N. P. (1996). Dimensions of stress among university faculty: Factor analytic results from a national study. *Research in Higher Education, 24*, 266–286.

King, A. (1992). Facilitating elaborative learning through guided student-generated questioning. *Educational Psychologist, 27*, 111–126.

Kranier, K. (2001). Teachers' growth is more than the growth of individual teachers: The case of Gisela. In F. L. Lin and T. J. Cooney (Eds.), *Making sense of mathematics teacher education* (pp. 271-293). Kluwer, Dordrecht, Netherland.

National Council of Teachers of Mathematics. (2007). *Mathematics teaching today: Improving practice, improving student learning.* Reston, VA: Author.

National Council of Teachers of Mathematics. (2000). *Principles and standards for school mathematics*. Reston, VA: Author.

Paulsen, M. B., & Feldman, K. A. (1995). Toward a reconceptualization of scholarship: A human action system with functional imperatives. *Journal of Higher Education, 66*, 615–640.

Samuelowicz, K., & Bain, J. D. (1992). Conceptions of teaching held by academic teachers. *Higher Education, 24*, 93–111.

Webster-Wright, A. (2009). Reframing professional development through understanding authentic professional learning. *Review of Educational Research, 79*, 702–739.

Wenger, E. (1998). *Communities of practice: Learning, meaning, and identity.* New York: Cambridge University Press.

Endnotes

The research reported in this article was supported by the National Science Foundation's Early Career Development Program DRL ROLE 0238319. All opinions are the responsibility of the author and do necessarily reflect the views of the Foundation.

The author also extends thanks to Andrea Norman for her assistance in compiling the video and survey data for this study.

Julie Cwikla is an associate professor of mathematics education at the University of Southern Mississippi Gulf Coast. She is Director of Project WetKids (www.projectwetkids.net), an out-of-school STEM program, and 2003 recipient of the National Science Foundation's Early Career Award.

Arbaugh, F., Lannin, J., Jones, D. L., and Barker, D.
AMTE Monograph 7
Mathematics Teaching: Putting Research into Practice at All Levels
© 2010, pp. 91–108

7

Textbook-Specific Professional Development: Impacting Teachers' Knowledge and Views

Fran Arbaugh
The Pennsylvania State University

John K. Lannin
University of Missouri

Dustin L. Jones
Sam Houston State University

David Barker
Illinois State University

Mathematics teachers experience many forms of professional development. This chapter describes a professional development project designed to support the implementation of a secondary problems-based textbook series. It reports findings from a study assessing the impact of a textbook-specific professional development on secondary mathematics teachers' knowledge and views about the textbook series, student learning, and teaching. Findings suggest that textbook-specific professional development can serve as a vehicle for simultaneous change in teacher views of instructional practices and student learning.

The adoption of a problems-based, reform-oriented textbook often requires learning a new way of teaching mathematics. For teachers who are quite skilled at a "teacher demonstrates–students practice" model of mathematics instruction, the transition to pedagogy that supports a more investigative way of learning can be difficult. Thus mathematics education literature has called for professional development designed to support teachers as they learn to implement a problems-based textbook (see, for example, Romberg, 1997). The community has heeded this call and has begun to design professional development focusing on the implementation of specific problems-based textbook series.

Knowing how teachers respond to professional development is important for understanding how to support them as they learn to use the text materials. Evidence suggests simply using a problems-based textbook may not be enough to support necessary changes in knowledge, beliefs, and instruction. Prawat (1992) found that one teacher's use of a new problems-based textbook resulted in changes in instructional practices, but did not result in a parallel change in the teacher's beliefs, both of which are necessary for sustaining an innovation. Arbaugh, Lannin, Jones, and Park-Rogers (2006) found that the number of years of teaching with a specific text series did not strongly correlate with the level of quality of the teachers' lessons using the series.

Research literature helps professional developers think deeply about designing and implementing worthwhile learning experiences for in-service teachers. But missing from this literature are the teachers' voices. In what ways do they experience and assess "effective" professional development? As Scribner (1999) notes, "Understanding how teachers experience professional learning is vital to create valuable (and valued) professional learning experiences. Few studies have examined this phenomenon in depth" (p. 231).

Further, few studies address professional development programs designed specifically for addressing issues arising when a particular mathematics textbook series is used. Existing research of this type focuses on the elementary and middle grades (See, for example, Grant & Kline, 2000; McDuffie & Mather, 2009; Pligge, Kent, & Spence, 2000).

A case of textbook-specific professional development occurred in Dickinson Public Schools [pseudonym] from 2002–

2005. Dickinson began using *Problems-Based Math* [pseudonym] with a small percentage of students (5%) and a few teachers in grades 9–12 during the 1996–97 school year. By 2001–02, more teachers and students (now in grades 8–12) were using this text series as their primary texts for high school mathematics. At that point, the first author of this chapter and the Dickinson 6–12 mathematics coordinator [Anonymous] received funding from the National Science Foundation (Arbaugh & Anonymous, 2002) to provide 200 hours of textbook series-specific professional development over two years for the teachers in grades 8–12.

As the professional development program concluded, the professional development providers wondered what teachers learned from the experience. During the three-year professional development project, formative assessment data had been gathered, allowing the adjustment of the original design as needed. However, the professional development providers wanted to understand the participating teachers' reactions to the program as a whole. Thus, delving into teachers' perceptions of project activities in this project became a goal. Understanding teachers' perceptions of professional development can guide the design of programs to best support the differing needs of teachers. Studying the impact of the professional development project from the perspective of the teachers reveals particular experiences they valued as they learned to use the text series.

Context and Nature of the Textbook-Specific Professional Development Project

Dickinson is a mid-western United States school district of over 16,000 students, with over 5800 students and approximately 50 mathematics teachers in grades 8–12. Since 1996, Dickinson has offered integrated mathematics courses using the *Problems-Based Math* textbook series to its students in grades 8–12. By 2005, approximately 50% of 8–12 students and 80% of 8–12 mathematics teachers were using the texts in mathematics classrooms. Of the 40 mathematics teachers using the series at that time, 35 had participated for three years in a textbook-specific professional development project titled "Getting to the Core."

"Getting to the Core" engaged teachers in two main components: summer workshops and academic-year study groups. Workshop content was organized two ways. One of the text series authors facilitated at least half of the workshop sessions, with the content of those sessions focusing on:

- The spiraling nature of mathematical topics through the four textbooks in the series (i.e., tracing the treatment of matrices through the four textbooks);
- The scope and sequence of the materials;
- The instructional sequence of launch, explore, share and summarize, and apply;
- Expectations for student mastery; and
- Research conducted around the text curriculum.

During the other half of the workshop sessions, project staff provided professional development organized around the needs of the participants. Those sessions included, but were not limited to, activities such as:

- Teachers working in course-alike groups (i.e., all of the textbook 1 teachers) to discuss implementation of specific lessons;
- Teachers making explicit connections between text content and the state curriculum guide; and
- Teachers working in school-based groups to develop "Parent Nights" that would facilitate communication with families of their students.

Participation in the summer institutes component supported teachers as they thought about such pedagogical issues as assessment practices, collaborative group work, and the use of technology; the mathematics they would be teaching; and planning for the upcoming year.

The second component of the project involved teachers participating in study groups during the academic year. Participating teachers selected the focus of their study group consistent with the organizational underpinnings of the study group model. Project staff encouraged teachers to choose a topic that would allow them to closely examine classroom-based issues of *Problems-Based Math* implementation. Examples of study group topics included:

- Working through the textbook as students;
- Designing and conducting an inquiry regarding students' perceptions of group work;
- Delineating a scope and sequence for a particular course, including planning common assessments;
- Observing group members' instruction and then debriefing;
- Participating in a modified version of lesson study; and
- Learning about how to support students' reading in mathematics class.

The study group meetings supported the teachers' reflections on implementation as well as planning for future lessons. In addition, these meetings contributed to the teachers' expanding knowledge base with regards to the textbook series. The combination of summer workshops and academic year study groups provided on-going, connected, and relevant professional development for the teachers.

Teachers' Reactions to the Professional Development Project

Approximately 35 teachers participated across the three years of the project. All teachers were compensated for the professional development that occurred outside of contract hours (some study group meetings were held during school hours. In that case, the project paid for substitutes). All teachers participated in data collection activities, including the interviews used for this investigation. Specifically, analysis for this investigation involved interviews from the 26 teachers who attended more than 75% of the available professional development.

The question guiding the analysis was, "In what areas did the teachers report that the professional development had the most influence?" This investigation was conducted as a phenomenography – a research method that uses interview as its primary data source and depends on self-report. "Phenomenography investigates the qualitatively *different* [emphasis added] ways in which people experience or think about various phenomenon" (Marton, 1996, p. 31). This theoretical view has great potential for the study of learning and its context because individuals experience and interpret learning environments in multiple ways. The goal is to understand people's perspectives of a phenomenon. A phenomenographer seeks to "uncover all of the understandings people have of specific phenomenon and to sort them into conceptual categories" (Marton, 1996, p. 32), and "these categorizations are the primary outcomes of phenomenographic research" (Marton, 1996, p. 33).

The answer to the question posed was three-pronged; the teachers believed that the professional development most influenced: their knowledge and beliefs about the textbook series; their instructional practices; and their views about student learning. In the following sections, details of these findings are presented.

Knowledge and Beliefs about the Textbook Series

Problems-Based Math was very different from both the textbooks that the project teachers used as learners of mathematics and those previously used when they taught. Thus, many project teachers reported that participating in "Getting to the Core" facilitated their learning about the design and philosophy of the textbook series itself. They reported completing the professional development project more knowledgeable about articulation within and across courses and the research base that exists with regard to these materials. In addition, teachers reported that the professional development had an influence on their beliefs and attitudes about using the textbook series with high school mathematics students.

Articulation within and across courses. Very few teachers in the project had teaching experiences in each of the four textbooks contained in the textbook series; most of the teachers had previously taught using only one or two of the textbooks. Consequently, most teachers had little knowledge of course content outside of their own teaching experiences. To increase teacher knowledge in this area, the professional development included opportunities for the teachers to examine strands of content across all four books in the series. This professional development activity appeared to influence the teachers' understanding of mathematical content articulation across the courses. For example, Ms. Williams reported:

> I think I am getting a better feel for the flow of the curriculum, the sequence of it. I just started teaching [from the third textbook]. I had not taught [from the first and second textbooks] and through the workshops, I have gotten to see the material and how it grows in each course and how it builds.... I mean it has been invaluable to me to look at the courses that are beyond us or prior to us to make sure that we understand where our kids have come from and where they are going to.

Ms. Nelson, a teacher in one of the three junior high schools (grades 8 and 9) in Dickinson, reported a similar gain in knowledge, but about courses that her students would take after completing her course:

> I teach [from the first textbook] and I can now see where it goes to [the rest of the series]....The first couple of years at my school, we all taught [from the first textbook]; we were just so involved in teaching that content. We did not know where the kids were going with it.

The teachers also spent time in professional development sessions engaged in discussing different components of a lesson. This appears to be time well-spent, as articulated by Ms. Kay:

> I understand better what all the parts of a lesson are. I understand how those parts fit together and what their purpose is a lot better than I did before the project.

Beliefs and attitudes about the textbook series. At the beginning of the professional development project, many teachers expressed strong opinions about the value of *Problems-Based Math* as a mathematics textbook series. Some teachers began the project "on board" with the problems-based approach of the series. Other teachers were less than enthusiastic about how different this approach was from their prior experiences. Although not consistent across all of the teachers in the professional development project, many teachers reported that the professional development had an impact on their beliefs and attitudes about the textbook series. For example, Ms. Grey stated:

> It is a combination of going through the professional development project and teaching it again that I see more value in this textbook series or a reform textbook series. I can see how when kids construct their knowledge that there is so much more growth there and I have become a proponent of it.... Now that I have taught it and been immersed in the [professional development] project, I realize how truly deep the thinking is that students do and I have really become a believer.

Ms. Long also spoke of having a shift in beliefs about the textbook series:

> I was not an advocate at all because I did not see the purpose of the mathematics in the first book. Now I realize that it is just a foundation for future content. I was frustrated because

> I was looking at the end product of my year and not the end product of the four-year process.

Teachers' Views of Changes in Instructional Practices

Teachers reported that this professional development experience influenced the ways in which they organized their classrooms, both pedagogically and in terms of managing classroom routines. Thus, the teachers viewed the professional development as a catalyst for changing instructional practices, as discussed below.

Transition to a more student-centered style. Many of the teachers claimed to have moved to a more student-centered pedagogical style as a result of participating in this professional development experience. Overall, teachers spoke about coming to the realization that they needed to, as Ms. Taylor described, "relinquish control of the class to the kids and [move] more from standing up and telling them to letting them work through it." She goes on to credit the professional development for this shift: "The big thing [the facilitators] did was help push me towards the kids doing it and not me doing it."

Other teachers echoed Ms. Taylor's description regarding the transition they had made toward a more student-centered classroom:

> Ms. McGraw: "I try not to lecture and tell kids everything anymore—they get on computers; we learn things from the program that the curriculum has.... I try to step out of the scene more and look for trouble spots and holes."

> Ms. Jefferson: "I have gone from standing in front of the room, pontificating, to almost never talking unless the kids force me to."

Perhaps Ms. Byer gave the most eloquent account of this shift to a more student-centered classroom:

> The point was to turn more than ever over to the kids rather than being more teacher-centered and in the front. I had relinquished some of that when I started teaching *Problems-Based Math*, but last year I decided, okay, I am really going to let go and see. I think the teacher in me felt that I had to keep [moving on in the textbook] because I had to accomplish so much in each period of time. Last year, I realized I wasn't covering the same amount of material [I had in past years]. But the kids were able to do more than I had ever seen them able to do before. Before I only watched them do what I had done. When I turned more over to them to think, I noticed that they were able to do math and think mathematically.... They were better thinkers because it had been left on their table to do.

Specifically, the teachers spoke of qualitatively different ways that their classroom practices were becoming more student-centered: they were becoming less rigid, becoming more patient, allowing students to struggle more with the mathematics, and listening more to students' mathematical thinking.

Assessing, collecting, and distributing student work. While leading one of the professional development sessions, Sarah Long, a long-time user of *Problems-Based Math* from another district, shared strategies she used to organize her classroom. Some teachers in the project tried the strategies and reported that they found them helpful in their classrooms. Among other strategies, Sarah shared that she had students correct their homework in red pen, so that she could then assess what students had completed at home versus what revisions they made to their work during a homework review. Ms. Taylor particularly found this strategy useful:

> [Correcting] homework with a red pen.... I like that because before I found that I was not going over homework very much or very well at all. I was asking if there were questions. There were rarely any questions because we had such a large number [of problems] and they hadn't done anything with it.

In addition, Ms. Taylor also commented on a strategy that Sarah shared regarding teachers assisting students during group work:

> She walks around the room all the time with the dry erase board in her hand. So, when the kids have a question and you want to pick up their pencil and write on their paper … she just writes on her dry erase board and she shows them and kind of gets them started. Then she erases it and goes on. That is a great idea.

More frequently, teachers spoke of the impact the professional development had on their commitment to and organization of group work. For example, Ms. Kellog stated:

> I have committed to sitting [sic] kids in groups, in all of my classes. In my *Problems-Based Math* classes, I am somewhat good at utilizing that seating arrangement.… So I am trying to use those ideas in all of my classes.

Ms. Hamm summed up the sentiments of the teachers when she said,

> Some of the things that I have used: the big thing is the classroom management. That was the hardest thing for me when I first did this, because I was used to straight rows—everybody in the row, everybody does their own thing. And to go from a sterile quiet type room to mass chaos, the first year was really difficult but I think I progressed from three.… Secondly, about dividing up the groups—I do that differently now. I do that totally randomly. I just use popsicle sticks and I draw names and I tell the kids if you don't like who you are with today, tomorrow you are going to be with somebody different so you know, get over it so you can get your work done. Whereas before I tried to place them in groups and I tried to arrange them.

All of these teachers appreciated that the professional development design included attention to helping them learn strategies for managing new classroom routines.

Teachers' Views of Student Learning

The participating teachers were well schooled in the "teacher demonstrates, students practice to mastery" model of mathematics instruction. Through the project, the teachers reported that they were better able to understand that implementing *Problems-Based Math* is different than what they were used to and that student learning "looked" different than before. For example, Ms. McGraw reported:

> Mastery does not always come when we think it should. In a traditional program, you introduce, quiz, master, move on. And this textbook's mentality is that mastery may not come until January or May. In this textbook, my thought is, it is circular—be patient. The next lesson will introduce something new, and still work on [what was in the previous lesson] and by the time you get to the end, come full circle, then, by golly, they got it. And it has taken me over time to see that—I will say that this project has helped me learn that.

Other teachers reported that they better understood the instructional model needed to use these materials to their fullest potential, as described by Ms. Rose:

> [The project] has given me a better understanding, a better knowledge, of what [the authors] are trying to do. I try to make sure that I am the coach—the students are pretty much doing the investigating. I am just kind of directing them the right way, instead of what I used to do, which was just telling them everything.

Ms. Casey agreed:

> I think I am more aware of what is going on, and trying to make sure that *Problems-Based Math* is implemented in the

> way it was intended to be: working in groups, investigative approach, giving [the students] more of a chance to mess around in the material.

Teachers also spoke about how the professional development had helped them be more patient with the ways that students learn using *Problems-Based Math*. Two teachers spoke about developing the ability to allow their students to grapple with the mathematics:

> Ms. Nelson: "[One of the] changes I made this year [is] my patience level with the kids … it is okay that the kids are struggling. I am getting better at not helping them as much, asking them the questions and getting them to think about it."

> Ms. Rose: "It is more of an opportunity for them to struggle and learn rather than for me just to tell them what to do. I feel like they probably learn a little bit more by struggling through it themselves."

In addition, Ms. Hamm reported that she had become more conscious of listening to her students' mathematical ideas:

> Well, I think that it [the project] has really helped me become a better teacher as far as seeing it form [sic] the perspective of the child a little more. The thing that has really been kind of difficult for me is being able to back away from my traditional type of mathematics that I have always taught before. So it has been a transition for me. But I think that it has gone pretty well, and I think that because of this [project] I have progressed a lot.

Discussion

"Getting to the Core" was designed around two themes: supporting teachers to build their knowledge about the textbook series; and facilitating discussion with regard to implementing the series in their classrooms. The analysis of teacher interviews indicated that the teachers viewed the impact of the professional development as occurring in three areas: teachers' knowledge about the textbook series; teachers' views of their instructional practices; and teachers' views of student learning.

Also, these three areas appeared to influence each other. As represented in Figure 1, some teachers described how the professional development and the discussions around the text series helped them better understand the goals of individual lessons and thus see the manner in which *Problems-Based Math* develops overarching concepts by returning to ideas through the use of a variety of problem solving situations. This realization of the difference in goals of lessons appeared to influence the teachers in considering learning as developing a deeper understanding of mathematical ideas over time through examining how ideas could and could not be applied to various problems. Further, this new knowledge facilitated a change in the view of their roles as teachers—away from a dispenser of knowledge towards a view as an active participant in guiding student learning.

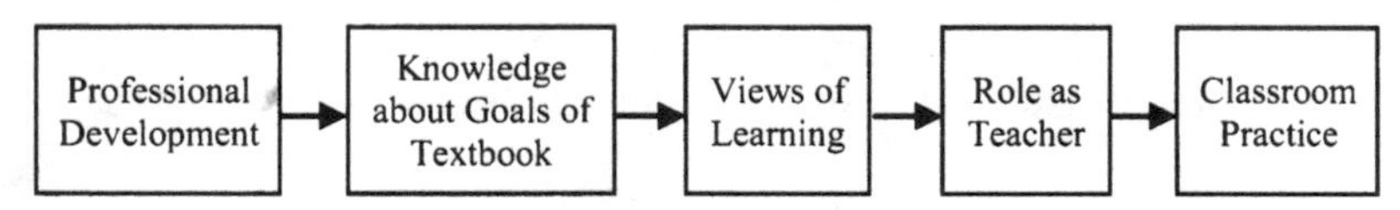

Figure 1. Influence of professional development: Path 1

However, Ms. Byers (quoted earlier in this chapter), who had two years of experience with *Problems-Based Math* at the beginning of the project, described a different sequence. The professional development influenced her thinking about her role as a teacher, which in turn influenced her classroom practices and her views of how her students learned (see Figure 2).

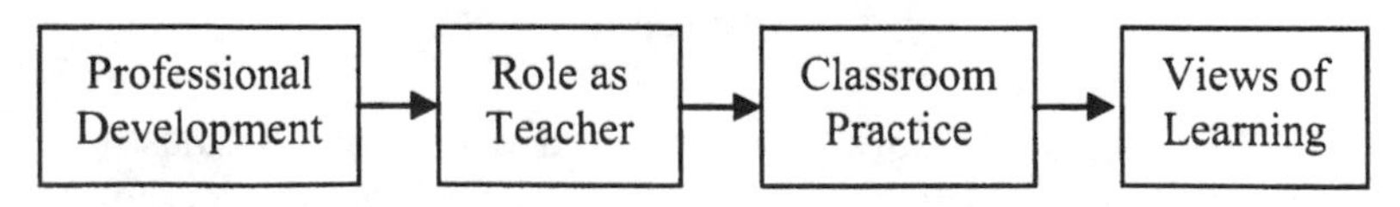

Figure 2. Influence of professional development: Path 2

Figure 3 conceptualizes how this experience of textbook-specific professional development appeared to influence teachers who participated in the project activities. Teachers reported that they were directly influenced in one or all of the areas shown due to their participation in the professional development. Our findings support the conceptualization of a type of professional development that focuses both on student learning and instructional practices, but through the lens of learning about a specific textbook series.

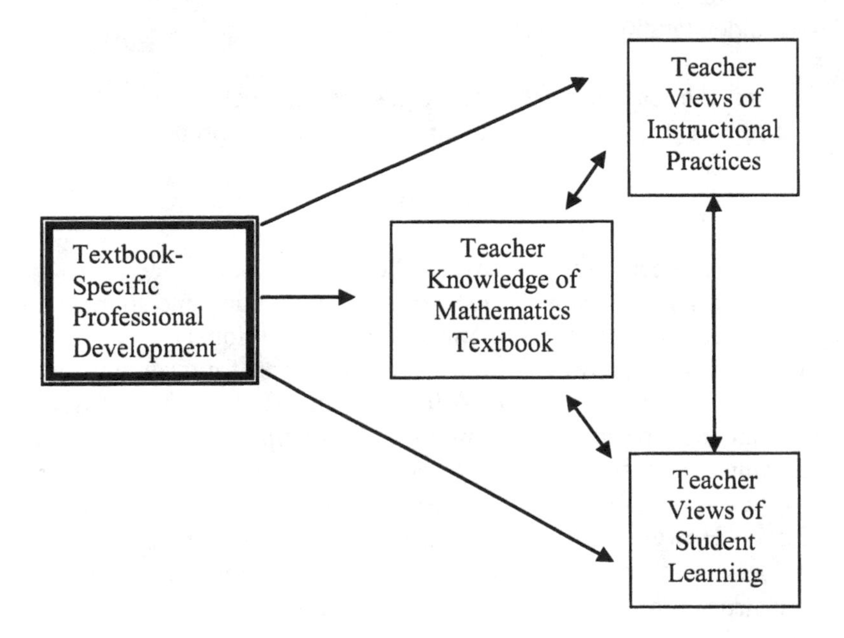

Figure 3. The influences of textbook-specific professional development

By initially focusing teachers' attention on learning more about the textbook series, the professional development engaged them in specific discussions with regard to how students learn mathematics as represented in *How People Learn: Brain, Mind, Experience, and School* (National Research Council, 2000). Issues such as mastery, when mastery occurs with these materials, and what mastery looks like when using a problems-based approach were addressed. The teachers were able to internalize the type of problems-based approach described by D'Ambrosio (2003):

> Proponents of teaching mathematics through problem solving base their pedagogy on the notion that students who confront problematic situations use their existing knowledge to solve those problems, and in the process of solving the problems, they construct new knowledge and new understanding. (p. 48)

Further, the teachers were able to discuss instructional practices that support learning through a problems-based approach.

Conclusion

This project led us to consider the importance of how curricular goals align with views of student learning and support for instructional practices. It appears that professional development must provide simultaneous support for change in these three intertwined areas. A focus on instructional practices without encouraging an examination of the impact of student learning and an adjustment of teacher views of the goals of instructional materials may be insufficient for promoting teacher change. However, textbook-specific professional development can serve as a vehicle for simultaneous change in teacher views of instructional practices and student learning.

References

Arbaugh, F., Lannin, J., Jones, D. L., & Park-Rogers, M. (2006). Examining instructional practices in *Core-Plus* lessons: Implications for professional development. *Journal of Mathematics Teacher Education, 9*, 517–550.

Arbaugh, F., & Anonymous (2002). Getting to the core: Supporting teachers' implementation of *Problem Solving*

Math [a pseudonym]. National Science Foundation funded Pilot Local Systemic Change project (ESI-0138556).

D'Ambrosio, B. S. (2003). Teaching mathematics through problem solving: A historical perspective. In H. L. Schoen (Ed.), *Teaching mathematics through problem solving: Grades 6–12* (pp. 39–52). Reston, VA: National Council of Teachers of Mathematics.

Grant, T. J., & Kline, K. (2000, April). *Understanding teachers' changing beliefs and practice while implementing a reform curriculum.* Paper presented at the American Educational Research Association, New Orleans.

Marton, F. (1996). Phenomenography – a research approach to investigating different understandings of reality. *Journal of Thought, 21,* 28–49.

McDuffie, A. R., & Mather, M. (2009). Middle school mathematics teachers' use of curricular reasoning in a collaborative professional development project. In J. T. Remillard, B. A. Herbel-Eisemann, & G. M. Lloyd (Eds.), *Mathematics teachers at work: Connecting curriculum materials and classroom instruction* (pp. 302–320). New York: Routledge.

National Research Council. (2000). *How people learn: Brain, mind, experience, and school.* Washington, DC: National Academy Press.

Pligge, M. A., Kent, L. B., & Spence, M. S. (2000, April). *Examining teacher change with the context of mathematics curriculum reform: Views from middle school teachers.* Paper presented at the American Educational Research Association, New Orleans.

Prawat, R. S. (1992). Are changes in views about mathematics teaching sufficient? The case of a fifth-grade teacher. *Elementary School Journal, 93*, 195-211.

Romberg, T. (1997). Mathematics in context: Impact on teachers. In B. Nelson & E. Fennema (Eds.), *Mathematics teachers in transition* (pp. 357–380). Mahwah, NJ: Lawrence Erlbaum Associates.

Scribner, J. P. (1999). Teacher efficacy and teacher professional learning: Implications for school leaders. *Journal of School Leadership, 9,* 209–234.

Fran Arbaugh is an associate professor of mathematics education at The Pennsylvania State University where she is involved in the preparation of undergraduate and graduate mathematics education students. She studies contexts for mathematics teacher learning as well as the development of mathematics teacher knowledge. She can be reached at arbaugh@psu.edu

John Lannin is an associate professor of mathematics education at the University of Missouri where he serves as Director of Elementary Education. He studies student mathematical reasoning and the development of mathematics teacher knowledge. He can be reached at lanninj@missouri.edu

Dustin L. Jones is an assistant professor of mathematics education at Sam Houston State University. He is interested in the design and impact of curriculum materials on instruction and student learning. He can be reached at dljones@shsu.edu

David Barker is an assistant professor of mathematics education at Illinois State University and is involved in the preparation of secondary teachers. His research interests include pedagogical content knowledge and students' use of representation. He can be reached at dbarker@ilstu.edu

Santagata, R., and van Es, E. A.
AMTE Monograph 7
Mathematics Teaching: Putting Research into Practice at All Levels
© 2010, pp. 109–123

8

Disciplined Analysis of Mathematics Teaching as a Routine of Practice

Rossella Santagata
University of California, Irvine

Elizabeth A. van Es
University of California, Irvine

This chapter describes efforts to design a preservice mathematics teacher education course focused on helping prospective teachers develop routine practices for learning from the analysis of teaching. Literature on expert teacher cognition as it relates to learning from teaching is briefly reviewed. A framework for the analysis of classroom lessons is introduced along with a description of proposed orientations and skills that preservice teachers need to implement the framework effectively. Sample resources and activities created to develop these orientations and skills are presented. The conclusion discusses issues arising through the design and enactment of the course and implications for teacher educators engaged in or planning similar experiences.

The use of portfolio assessments in preservice teacher education has become more prominent in the last decade. These portfolios are used as evidence for teaching competence. Such new forms of assessment are rooted in research that argues for teacher education to help prospective teachers "become serious learners in and around their practice" (Ball & Cohen, 1999). This perspective suggests that a critical component of teacher preparation is learning in the contexts of one's work. According to Ball and Cohen (1999), teachers need "to attend to and learn about individual students' knowledge, ideas, and intentions...and to stand back from and analyze their own teaching, to ask and answer such questions as: What is working? What is not working? For whom are some things working or not working?"

More recently, Hiebert, Morris, Berk, and Jansen (2007) proposed that preservice teachers learn how to teach from studying teaching, arguing that future teachers need to be equipped with skills and knowledge to learn from their own teaching practice. Practice-based approaches to teacher learning have focused on developing knowledge and skills that allow for the implementation of such "high-leverage practices" (Hatch & Grossman, 2009, p. 76) as posing, solving, and sharing solutions to word problems or monitoring student work during seatwork (Kazemi, Lampert, & Ghousseini, 2007). This paper's authors contend that together with the instructional routines of practice, teachers need to develop knowledge and skills to routinely analyze and learn from their work. In other words, disciplined analysis of teaching should be part of teachers' repertoire of practices.

Efforts to create a preservice teacher education course that focuses precisely on helping preservice teachers develop routine practices for learning from teaching led to the Learning to Learn from Teaching project, begun in 2007. This project sought to develop frameworks for designing instruction and to conduct empirical research on the impact of learning from teaching experiences on preservice teachers' knowledge, beliefs, skills, and practices as they progress through a teacher education program and begin their teaching careers. The project involved development of two versions of the course: one for future elementary teachers and one for future secondary mathematics teachers.

Literature on expert teacher behavior as it relates to learning from teaching is briefly reviewed to introduce a framework for the analysis of classroom lessons at the center of the Learning to Learn from Teaching course. Following that is a description of practices that preservice teachers need to master to implement the framework effectively, including examples of activities used in the context of the course. Finally, issues arising through the design and enactment of the course are discussed for teacher educators engaged in or planning similar experiences.

Expert Teacher Behaviors

Expert teachers use evidence of student learning or difficulties to make decisions about instructional strategies they will use next, and in doing so they reason about the impact of specific teacher moves on student learning. Borko and Livingston (1989) revealed that expert teachers, when reflecting on their own lessons, selected classroom events that had an impact on the achievement of the lesson learning goals. Similarly, Berliner (2001) concluded that while novice teachers tend to adhere rigidly to lesson plans, expert teachers are more flexible, paying attention to student difficulties during instruction, reasoning about them, and making on-the-spot decisions in response.

The ability to attend to and reason about student thinking is referred to as "noticing" (Ainley & Luntley, 2007; Jacobs et al., 2010; Mason, 2002; van Es & Sherin, 2008). In a study of professional development focused on children's mathematical thinking, Jacobs and colleagues (2010) found that experienced teachers who engaged in over four years of sustained professional development possessed better noticing skills and were better able to respond to student ideas than teachers with fewer years of the same training. It is neither reasonable nor realistic to expect preservice teachers to develop expert skills in a teacher credential program. On the other hand, given recent calls for mathematics teachers to listen carefully to student ideas and use those ideas to inform instructional decisions, it is particularly important to help preservice teachers develop both beginning analytic skills for examining their teaching and an orientation to teaching as learning.

The Lesson Analysis Framework

The *Lesson Analysis Framework* (Hiebert et al., 2007; Santagata, Zannoni, & Stigler, 2007), drawn from the novice-expert literature summarized above, is designed to assist teachers in the analysis of teaching by focusing on the connections among different aspects of teaching and using evidence of student learning to guide decision-making (Borko et al., 2008; Davis, 2006). The framework centers the analysis of teaching on classroom lessons representing natural units in the process of teaching. It consists of a series of questions that guide teachers through a lesson analysis process.

The first question asks teachers to analyze the lesson learning goals: What are the main ideas that students are supposed to understand through this lesson? Teachers then move to the analysis of student learning by attending to the following questions: Did the students make progress toward the learning goals? What evidence do we have that students made progress? What evidence do we have that students did not make progress? What evidence are we missing? Analyzing the particulars of student learning and understanding as evidenced in the lesson leads teachers to the next question, focused on the impact of teachers' decisions on student learning: Which instructional strategies supported students' progress toward the learning goals and which did not?

Finally, building on the analysis of the cause-effect relationship between teaching and learning, teachers are asked: What alternative strategies could the teacher use? How do you expect these strategies to impact students' progress toward the lesson learning goals? If any evidence of student learning is missing, how could the teacher collect such evidence? These last questions are important because they connect the analysis of practice with action in teaching. Research investigating the use of the Lesson Analysis Framework on teacher learning has documented improvements in preservice teachers' abilities to analyze teaching (Santagata et al., 2007; Santagata & Angelici, in press).

Orientations and Skills Necessary to Implement the Lesson Analysis Framework

In order to effectively answer the questions described above, preservice teachers need to develop certain orientations and master several skills. These fall into three main categories: awareness, knowledge and analysis skills, and planning and enactment. First, to be able to effectively implement the Lesson Analysis Framework, preservice teachers need an awareness of the importance and usefulness of a disciplined analysis of practice (Rodgers, 2002). They need to appreciate the value of a teaching approach built on student ideas, and they need a realization of the complexity of student thinking about mathematical ideas.

Second, the orientations must be coupled with specific knowledge and analysis skills, including the ability to attend to what students are doing or saying in a lesson and to draw inferences or make hypotheses based on mathematical understanding. Preservice teachers must also have knowledge of strategies that assist in making students' thinking visible. This knowledge involves learning to identify and examine such key routines as effective questioning, designing open-ended mathematical problems, monitoring student work, and establishing a classroom discourse community. These practices are a focus of analysis because they align with research on teaching mathematics for understanding (Hiebert et al., 1997).

Finally, planning and enactment skills are crucial to move preservice teachers to analyze their own teaching, not simply that of others. The first needed skill is the ability to generate alternative strategies and justify them in terms of potential impact on student learning. This skill can be practiced in the context of analyzing the teaching of others first and then applying what is learned to one's own teaching. The second and third skills deal with the ability to plan teaching and enact instructional practices that make student thinking visible.

The orientations and skills described above constitute the learning objectives for the Learning to Learn from Teaching course. Both versions of the course include activities tailored to levels matching the preservice teachers' credentials and target developing these orientations and skills. What follows is a sample of course activities that focus on two sets of skills.

Skill set 1: Analysis of student learning and thinking. Preservice teachers engaged in analysis of student thinking by viewing video clips showing at least two different student ideas or solution strategies for the same problem. Preservice teachers solved the problem before they saw the clips and shared the different strategies they generated. They then viewed the corresponding clips with an accompanying transcript and were asked to identify different student ideas, prompted by such questions as what students appeared to understand mathematically, what confusions or difficulties they seemed to have, and from where might these originate. In supporting hypotheses about student thinking, preservice teachers were encouraged to highlight particular evidence in the transcript and points in the video segment.

The class then reviewed the same clip and examined how the teacher made student thinking visible, focusing on what the teacher did to unveil different student ideas and confusions. Three particular teaching strategies were analyzed: how the task was designed to elicit student talk and images of student work, how the questions asked by the teacher and students developed the ideas discussed, and how the teacher monitored student learning and responded to individual student ideas. The result of this analysis was a generated list of particular teaching strategies for eliciting and probing student thinking. This list became a resource for subsequent activities focused on the planning and enactment of mathematics lessons that make student thinking visible.

This analytic process focused initially on several 5- to 10-minute video segments, allowing a fine-grained examination of the interactions among the students and teacher. More complete lessons were then analyzed to examine the progression of student thinking and to see a range of teaching strategies for eliciting student thinking over the course of a lesson. Each cycle of analysis resulted in more elaboration of the generated list of effective teaching practices for making student thinking visible.

Skill set 2: Planning for and enacting routines of practice to elicit student thinking. To develop these skills, preservice teachers worked in grade-level groups and planned together for a discussion involving four to five students around the solution of an open ended problem. To plan the discussion, they used a template developed by the Mathematics Methods Planning Group at the University of Michigan (2006). This template includes seven sections that guide the planning of the various parts of a discussion: (1) choosing the problem; (2) anticipating student thinking; (3) setting up the problem; (4) monitoring student work, (5) launching the discussion; (6) orchestrating the discussion; and (7) concluding the discussion. Each section includes guiding questions that encourage preservice teachers to be specific about strategies used to make student thinking visible.

Course participants then enacted plans independently with four or five students from their fieldwork class. They videotaped and transcribed the discussions (approximately 30 minutes long) and collected student work, then wrote a report supplementing the group plan and the transcript with a reflection and analysis of the effectiveness of their teaching supported by evidence from the transcript and from student work. Student work and analyses of teaching were also shared with other preservice teachers from the same group.

The activity ended with a poster presentation during which groups presented the problem they posed, showed student solutions (from the least to the most sophisticated), and summarized what they learned from the experience and steps for improving teaching. Other activities followed a similar structure: pre-service teachers were given opportunities to analyze teaching independently, to work in groups, and to share their analyses in the larger group. This allowed for the development of a shared language (supported by concrete representations) to describe and reason about teaching.

The next section discusses several issues that arose in the process of designing and enacting the Learning to Learn from Teaching course.

Design Considerations

Video Materials

The first issue to consider is the selection of video materials for analysis. The authors identified a variety of published materials representing mathematics teaching and learning. Selected video materials (see Table 1) captured the overall lesson structure, a range of student thinking and ideas, and a range of teaching strategies that made student thinking visible.

Table 1
Video Resources for Analyzing Teaching

Learning Goal	Grade Level	Video Case Materials
Individual Student Thinking	Elementary	*Cognitively Guided Instruction* (Carpenter, Fennema, Franke, Levi & Empson, 1999)
		IMAP: Integrating mathematics and pedagogy to illustrate children's reasoning (Philipp & Cabral, 2005)
	Secondary	*The Clinical Interview* (Walbert, 2001)
Small Group Discussions	Elementary	*Number and Operations: Building a System of Tens, Casebook and Facilitator's Guide* (Schifter, Bastable, & Russell, 1999)
	Secondary	*Learning and Teaching Linear Functions* (Seago, Mumme, & Branca, 2004)
		Principles of Learning Video Cases (Institute for Learning, 2007)
Edited Whole Class Excerpts	Elementary	*Cognitively Guided Instruction* (Carpenter, Fennema, Franke, Levi & Empson, 1999)

Learning Goal	Grade Level	Video Case Materials
Edited Whole Class Excerpts	Elementary	*Teaching Math: A Video Library* (Annenberg Media at learner.org)
	Secondary	*Connecting Mathematical Ideas* (Boaler & Humphreys, 2005)
		Teaching Math, A Video Library (Annenberg Media at learner.org)
Unedited Whole Class Lesson	Secondary	*Teaching Mathematics in Seven Countries: TIMSS Video Study* (1999).

When appropriate, additional frameworks such as the math-talk learning community and the task complexity frameworks (Huffered-Ackles, Fuson, & Sherin, 2004; Stein, Smith, Henningsen, & Silver, 2000) were used to scaffold the preservice teachers' analysis of teaching. Deliberate choices were made about whether to review the same clip multiple times or to show preservice teachers a range of videos of teaching. One of the great virtues of video is the ability to revisit a clip through multiple analytic lenses (i.e., using different frameworks to analyze the same clip). However, the preservice teachers wanted to see a range of examples of teaching and often questioned the value of viewing the same clip multiple times or watching the same teacher in different clips. The authors believe it is valuable to view some clips more than once, but it is also important for preservice teachers to see images of many teachers making student thinking visible and employing strategies to do so.

Balancing Images of Beginning and Expert Teaching

When choosing video recordings for modeling effective practice, teacher educators often opt for video of teachers that the education community has recognized as expert in a particular practice or for videos developed by teaching organizations and foundations to purposely portray expert practice. With similar intent, the authors chose to show expert teachers enacting strategies for making student thinking visible. The decision was motivated by the necessity to give student teachers opportunities to observe practices seldom seen during fieldwork (Feiman-Nemser & Buchmann, 1986). This choice had a positive impact in that course participants began to form concrete images of the desired practices and were able to critically reflect on fieldwork placements.

However, these videos only provided images of an ideal end product. That is not to say that the teaching examples had no weaknesses; in fact, course participants were encouraged to reflect on ways the video recorded lessons could be improved. What the videos did not illustrate is the trajectory a novice teacher might follow to master certain practices. One preservice teacher commented:

> I think it's good and I think it's the ideal. So I guess why not be exposed to an ideal.... [But] I mean I have no idea how that teacher we observed got to the level that she did with the discussion with the children who were so respectful of each other and really moved to ask the right questions to probe each other.

With this in mind, teacher educators should supplement video portraying expert teachers with video portraying novice teachers who have mastered some elements of a certain practice, but not others, or who have mastered a certain practice, but not consistently. This gives preservice teachers the opportunity to observe a level of teaching that is within their reach, while maintaining the ideal on the horizon.

Taking Ownership of Visible Thinking Teaching Strategies

A third design issue concerns helping pre-service teachers take ownership of these teaching practices. Pre-service teachers often encounter conflicting images of teaching in the field and at the university. While the Learning to Learn from Teaching course advocated making student thinking visible, not all preservice teachers were placed in classrooms where the cooperating teacher modeled such practices. Furthermore, many preservice teachers had not experienced such learning as students themselves. Thus, both their personal conceptions of teaching and the images they saw in the field challenged the teaching practices that the course promoted. This made it difficult for the preservice teachers to embrace and enact visible thinking teaching strategies. A response to these challenges would be to include more role-playing in the teacher education course (Grossman & McDonald, 2008). In addition, teacher preparation programs could develop close partnerships with cooperating teachers in order to develop a shared vision of mathematics teaching.

Preservice teachers enter teacher education programs with ideas about what they should learn. They want to learn how to plan a lesson along with strategies to manage students and deliver content to students. They are persistent in wanting to know "what to do." However, a core goal of the Learning to Learn from Teaching course involved helping preservice teachers realize that reflecting on teaching *is* part of teaching. While respecting the need for preservice teachers to learn specific strategies that they can put into practice, it is fundamental for teacher educators to explicitly challenge them to expand their vision of what it means to engage in the work of teaching to include disciplined analysis of practice.

Conclusion

This paper articulates a framework for helping preservice teachers develop knowledge and practices for inquiring into teaching. Hiebert and colleagues (2007) advocate that preservice teacher education is the time to develop such skills; however, little knowledge exists for how to design a curriculum for prospective teachers to accomplish this goal. The Learning to Learn from Teaching project contributes to this work by specifying learning goals for preservice teachers, by articulating the key skills they need, and by sharing strategies to help them learn to learn from teaching. Future research will investigate the effect this course has on preservice teachers as they progress through a teacher education program and begin teaching.

References

Ainley, J., & Luntley, M. (2007). The role of attention in expert classroom practice. *Journal of Mathematics Teacher Education, 10*, 3–22.

Annenberg Media, Learner.org. *Teaching Math: A Video Library.* Accessible at: www.learner.org

Ball, D. L., & Cohen, D. K. (1999). Developing practice, developing practitioners: Toward a practice-based theory of professional education. In G. Sykes and L. Darling-Hammond (Eds.), *Teaching as the learning profession: Handbook of policy and practice* (pp. 3–32). San Francisco: Jossey Bass.

Berliner, D. C. (2001). Learning about and learning from expert teachers. *International Journal of Educational Research, 35*, 463–482.

Boaler, J., & Humphreys, C. (2005). *Connecting mathematical ideas: Middle school video cases to support teaching and learning.* Portsmouth, NH: Heinemann.

Borko, H., Jacobs, J., Eiteljorg, E., & Pittman, M. E. (2008). Video as a tool for fostering productive discussions in mathematics professional development. *Teaching and Teacher Education, 24*, 417–436.

Borko, H., & Livingston, C. (1989). Cognition and improvisation: Differences in mathematics instruction by expert and novice teachers. *American Educational Research Journal, 26*, 473–498.

Carpenter, T. P., Fennema, E., Franke, M. L., Levi, L. W., & Empson, S. B. (1999). *Children's mathematics: Cognitively guided instruction.* Portsmouth, NH: Heinemann.

Davis, E. A. (2006). Characterizing productive reflection among preservice elementary teachers: Seeing what matters. *Teaching and Teacher Education, 22*, 281–301.

Feiman-Nemser, S., & Buchmann, M. (1986). The first year of teacher preparation: Transition to pedagogical thinking. *Journal of Curriculum Studies, 18*, 239–256.

Grossman, P., & McDonald, M. (2008). Back to the future: Directions for research in teaching and teacher education. *American Educational Research Journal, 45*, 184–205.

Hatch, T., & Grossman, P. (2009). Learning to look beyond the boundaries of representation: Using technology to examine teaching (Overview for a digital exhibition: Learning from the practice of teaching). *Journal of Teacher Education, 60*, 70–85.

Hiebert, J., Carpenter, T. P., Fennema, E., Fuson, K. C., Wearne, D., Murray, H., Human, P., & Olivier, A. (1997). *Making sense: Teaching and learning mathematics with understanding*. Portsmouth, NH: Heinemann.

Hiebert, J., Morris, A. K., Berk, D., & Jansen, A. (2007). Preparing teachers to learn from teaching. *Journal of Teacher Education, 58*, 47–61.

Hufferd-Ackles, K., Fuson, K. C., & Sherin, M. G. (2004). Describing levels and components of a math-talk learning community. *Journal for Research in Mathematics Education, 35*, 81–116.

Institute for Learning (2007). *Principles of learning: Study tools for educators* (Version 2.0) [CD-ROM]. Pittsburgh, PA: University of Pittsburgh.

Jacobs, V. R., Lamb, L. L. C., & Philipp, R. A. (2010). Professional noticing of children's mathematical thinking. *Journal of Research in Mathematics Education, 41*, 169-202.

Kazemi, E., Lampert, M., & Ghousseini, H. (2007). *Conceptualizing and using routines of practice in mathematics teaching to advance professional education.* Chicago: Spencer Foundation.

Mason, J. (2002). *Researching your own practice: The discipline of noticing*. London: RoutledgeFalmer.

Mathematics Methods Planning Group. Discussion planning framework. Retrieved from http://www-personal.umich.edu/~sleepl/ed411classdocs2006.htm

Philipp, R., & Cabral, C. (2005). *IMAP: Integrating mathematics and pedagogy to illustrate children's reasoning*. Pearson, NY: Merrill Prentice-Hall.

Rodgers, C. R. (2002). Seeing student learning: Teacher change and the role of reflection. *Harvard Educational Review, 72*, 230–253.

Santagata, R., Zannoni, C., & Stigler, J. W. (2007). The role of lesson analysis in pre-service teacher education: An empirical investigation of teacher learning from a virtual video-based field experience. *Journal of Mathematics Teacher Education, 10*, 123–140.

Santagata, R., & Angelici, G. (in press). Studying the impact of the lesson analysis framework on pre-service teachers' ability to reflect on videos of classroom teaching. *Journal of Teacher Education.*

Schifter, D., Bastable, V., & Russell S. J. (1999). *Number and operations: Building a system of tens, casebook and facilitator's guide*. White Plains, NY: Dale Seymour Publications.

Seago, N., Mumme, J., & Branca, N. (2004). *Learning and teaching linear functions: Video cases for mathematics professional development*. Portsmouth, NH: Heinemann.

Stein, M. K., Smith, M. S., Henningsen, M. A., & Silver, E. A. (2000). *Implementing standards-based mathematics instruction: A casebook for professional development*. New York: Teachers College Press.

Teaching Mathematics in Seven Countries: TIMSS Video Study (1999). Philadelphia, PA: Research for Better Schools.

van Es, E. A., & Sherin, M. G. (2008). Mathematics teachers' "learning to notice" in the context of a video club. *Teaching and Teacher Education, 24*, 244-276.

Walbert, D. (2001). *The Clinical Interview.* Retrieved from Learn NC Edition website: http://www.learnnc.org/lp/editions/pcmath/1.0

Endnote

This research was conducted with support in part by a fellowship from the Knowles Science Teaching Foundation to Elizabeth van Es. The opinions expressed are those of the authors and do not necessarily reflect the views of the supporting agency. The authors are listed alphabetically. They each contributed equally to the development of this research.

Rossella Santagata is an assistant professor in the Department of Education at the University of California, Irvine. Her research focuses on the use of video as a tool for mathematics teacher preparation and professional development. She can be contacted at r.santagata@uci.edu

Elizabeth van Es is an assistant professor in the Department of Education at the University of California, Irvine. Her research focuses on teacher noticing and uses of video in teacher education. She can be contacted at evanes@uci.edu

Suh, J., and Parker, J.
AMTE Monograph 7
Mathematics Teaching: Putting Research into Practice at All Levels

9

Developing Reflective Practitioners through Lesson Study with Preservice and Inservice Teachers

Jennifer Suh
George Mason University

Jana Parker
George Mason University

This case study describes pre-service teachers collaboratively planning and reflecting with cooperating teachers and other educators at their clinical site. Using lesson study as the professional development structure, preservice teachers worked with classroom teachers, resource specialists and mathematics educators while being immersed in authentic teaching situations that revealed complex pedagogical issues and factors impacting the teaching and learning of mathematics. Qualitative analysis of teacher interviews, reflections, classroom observations, and planning documents revealed several unique outcomes including developing mathematical knowledge for teaching through a reciprocal learning process; revealing specific gaps in mathematical knowledge for teaching among preservice teachers and increasing preservice teachers' awareness of the complexity of teaching and reflective practice. Finally, the study identifies specific critical norms for ensuring the success of lesson study among preservice and practicing teachers.

Contributing to the understanding of how lesson study supports preservice and inservice teachers in developing mathematical knowledge for teaching and reflective practice, this case study describes preservice teachers collaboratively planning and reflecting with cooperating teachers and other educators at clinical sites. Through engaging in lesson study with practicing teachers, preservice teachers were exposed to authentic pedagogical issues situated in the context of teaching using discussions of students' common misconceptions and factors related to specific student populations (i.e., English language learners and special needs students) that influenced instructional decisions and professional learning.

Research on Reflective Practice and Lesson Study

Ma (1999) stated that American teachers believe they need to know mathematics to plan lessons, whereas Chinese teachers think they can learn mathematics through planning lessons. Learning through collaborative planning, teaching, observing, and debriefing affords opportunities for teachers to reflect individually and collectively. Kolb (1984) described an experiential learning cycle that begins with a concrete experience and moves on to reflective observations about that experience, which in turn guide a stage of active experimentation continuing the cycle. Reflective practice is a cornerstone of best practices in teacher education that develops the analytical and inquiring disposition of teachers. Multiple opportunities for reflection are needed to build teachers' capacity for critical reflection. Some traditional methods of fostering reflection have been through journaling, video recording class sessions, and conferences with mentors or colleagues. Lesson study as a collaborative structure for developing reflection has drawn attention and become a catalyst for critical dialogue about mathematics teaching and learning among teachers (Lewis, Perry, & Murata, 2006).

A growing body of research (Chazan, et al., 1998; Fernandez & Yoshida, 2004; Roth & Tobin, 2004) supports collaborative inquiry in teaching and learning as a highly effective component of professional development for teachers. As a result, preservice and inservice teachers are urged to participate in lesson study (Fernandez & Yoshida, 2004) involving a cycle of collaboratively planning a research lesson, teaching and observing the lesson, reflecting on and revising the lesson, and repeating the cycle.

Lesson planning involves establishing lesson objectives, evaluating instructional materials, determining how to assess student understanding, reviewing one's understanding of mathematical concepts, and situating an instructional experience in the curriculum. All these tasks provide opportunity for individual and collective reflection on content and pedagogy and have the potential to deepen a teacher's mathematical knowledge for teaching and pedagogical content knowledge. However, preservice teachers have limited experience in planning lessons and are not adept at performing the task. Therefore, it is critical for mathematics educators to design experiences that allow preservice teachers to take advantage of the opportunities presented by lesson planning.

Benefits of collaborative lesson planning include exposure to multiple perspectives and new ideas that result from the pooling and sharing of experiences. Teachers engage in discussions about mathematics content involving mathematically accurate explanations that are comprehensible and useful for students and represent ideas clearly and precisely. Smith, Bill, and Hughes (2008) use the "Thinking Through a Lesson Protocol," which prompts teachers to think deeply about a specific lesson to be taught. Teachers use the protocol to move beyond structural components of lesson planning to a deeper consideration of how to advance students' mathematical understanding during the lesson.

Some research (Fernandez and Yoshida, 2004) on collaborative planning emphasizes using a lesson-plan format, particularly a four-column plan, including sections for instructional activities, anticipated student responses, teachers' proposed reactions to student responses, and assessment. Others, such as Hawbaker, Balong, Buckwalter, and Runyon (2001), describe a four-component method for planning: identifying the big ideas, analyzing areas of difficulty, creating strategies and supports, and evaluating the process. These researchers consider the role of the actual lesson plan when they describe the benefits of collaborative planning. And yet others (Chazan, et al., 1998; Walther-Thomas, 1997) note advantages resulting from collaborative planning without identifying a specific lesson plan format.

This paper explores how collaborative reflection through lesson study supported reflective practice and mathematical knowledge development for preservice and inservice teachers. Attention is given to detailing the process for other mathematics educators to use the collaborative learning structure in their work.

The Lesson Study Project

Using case study methodology, the researchers focused on five of 22 preservice teachers in a professional development school who participated in a yearlong internship while concurrently taking a university mathematics methods course meeting one day per week. The 3-credit methods course addressed mathematics content and pedagogy with a focus on designing mathematics lessons, using technology effectively in mathematics instruction, and assessing student learning through performance-based assessments. Preservice teachers in the course were required to plan, deliver and reflect on three lessons during the course of the semester. As a pilot project, the instructor (and one of the researchers) supervised five preservice teachers at a professional development school and created opportunities for the preservice teachers to participate in lesson study, collaborating on the research lesson design with cooperating teachers and other practicing educators.

Context for the Lesson Study

To begin collaborative planning, the instructor engaged both preservice and inservice teachers, teachers of English language learners, and special educators in planning a lesson for a diverse student population in a Title One school. Three preservice teachers placed in a primary internship participated in one lesson on subtraction with regrouping which took place in a second grade class, while two others joined a research lesson on decimal place value for a fifth grade class. Both groups of teachers used a four-column lesson plan format (see Figure 1) to structure the planning process. In this model, Column 1 outlines the flow of instruction in detail; Column 2 maps out anticipated student responses and solution strategies; Column 3 details teacher responses to differentiate for diverse learners; and Column 4 addresses assessing student understanding.

Title ______________ Content Area: ______________
Teacher name ______________ Grade Level ______ Lesson Study Date ______

OVERARCHING GOAL FOR THE LESSON STUDY	LESSON OBJECTIVES and STANDARDS		
IMPORTANT MATHEMATICS CONTENT BACKGROUND: Describe the important mathematical concepts related to this lesson that a (teacher/students) have as prior or future math concepts to (teach/learn)			
MATERIALS: List the texts, equipment, and other materials to be used by the students. List the materials, including equipment or technology used by the teacher in presenting the experiences			
Steps of the lesson: learning activities and key questions (and time allocation)	Anticipated Student Responses and solution strategies. (Potential Barriers & Misconceptions)	Teaching notes. DIFFERENTIATION: List adaptations for GT, ESOL, LD	Evidence of learning Evaluation points or assessments questions
LINK PRIOR KNOWLEDGE: Outline procedures for activating prior knowledge and student interest.			
INSTRUCTIONAL STRATEGIES: Outline what the teachers and students will do to Engage & Educate. Active learning tasks			
REFLECT and SUMMARIZE: Outline how you will close.			
EXTENSIONS/CONNECTIONS: What other lessons does this lesson connect to?			
REFLECTION: After the lesson, reflect on what went well and what didn't go well. Write changes you might implement the next time the lesson is taught			

Figure 1. Four-column lesson plan format modified from Lewis (2002).

The two groups created concept maps (see Figure 2) illustrating identified mathematical ideas related to the lesson. These tools allowed all to discuss students' prior knowledge and future mathematical building blocks, important vocabulary, and prerequisite knowledge necessary for students to access the lesson.

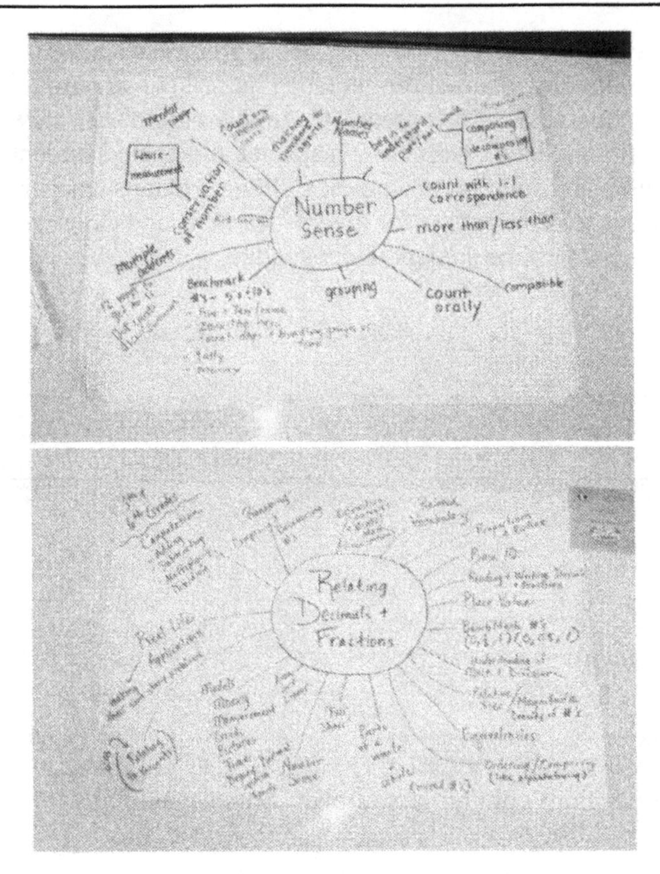

Figure 2. Mathematical concept maps from 2nd and 5th–6th grade groups.

The lesson study process involved four phases, spanning four two-hour after school sessions and one release day. It included collaborative planning, teaching and observation, a debriefing and reflecting phase, and refining and enhancing the lesson. In the collaborative planning phase, all participants collaborated on lesson planning during after school sessions. The teaching and observation phase and the debriefing phase, took place on the release day. The host, a cooperating teacher, taught the focus lesson while observers wrote individual reflections and notes. In the debriefing phase, the lesson study teams met again as a group to reflect on lesson design, the task, student engagement and learning, and future steps including revisions. In subsequent cycles, preservice teachers taught the revised lesson with observers watching and helping debrief the experience.

Some guiding questions crucial to the teaching and learning processes were: What is the important mathematical understanding that students need to learn? What are potential barriers and anticipated student responses? What conceptual supports and instructional strategies can best address our students' learning? How will we respond when students have difficulty? How will we know when each student has learned the mathematics?

Following the lesson study, teachers were asked to respond to two prompts: Describe your experience with lesson study in terms of personal and professional gains, challenges and "aha" moments; and, Which column of the four-column lesson plan was most helpful, and in what ways? Additionally, preservice teachers wrote a summative reflection of the process. Finally, two preservice teachers from each group were interviewed individually by researchers about the experiences.

Learning Outcomes from the Project

Using the constant comparative method (Strauss & Corbin, 1994), researchers identified common themes in the preservice teachers' written reflections. Data sources included transcribed interviews, reflective journal entries, classroom observations, and planning documents. The qualitative data were analyzed using open coding techniques and tested for themes and patterns. Emergent ideas were categorized into themes and crosschecked with teachers' comments and researchers' notes. Three key learning outcomes were recurrent in the analysis and are discussed in the following sections.

Developing mathematical knowledge for teaching through reciprocal learning. A reciprocal learning relationship among all participants was evident in the discourse that occurred during the planning phase. Different levels of mentoring and expertise were revealed as each contributed to the group's knowledge. Preservice teachers were mentored by experienced teachers and special educators, who shared their knowledge of potential barriers to learning, common misconceptions, and anticipated student responses acquired through years of training and experience working with diverse student populations. For example, when planning for a fifth and sixth grade lesson on decimal numbers, practicing teachers discussed common student misconceptions about decimal place value. A preservice teacher wrote the following journal entry after a planning session:

> I did not think that when some kids see a decimal number like .79 and .8, that they might think that .79 is greater because they are disregarding the decimal point and merely thinking 79 versus 8 as whole numbers. We had a lengthy discussion about how we can develop a lesson that would reveal this misconception. We decided that the Decimal Draw Game would bring this out by having students create the largest decimal numbers to win a game using the digits that they would roll [on a die]. I think this will be a great way [for] them to have to argue why one number is greater than the other.

In contrast, preservice teachers working with second grade teachers most frequently shared their knowledge of strategies, curriculum and technology tools. Initially, preservice teachers were passive; however, as the sharing continued, they developed confidence and felt their voices were validated. During a follow-up interview, one preservice teacher reported:

> In the beginning, I did not really participate because I was a little intimidated to be surrounded by so many teachers with years and years of experiences. I was not sure about in what way I could contribute to the planning of the lesson. But as the planning processes continued, I was encouraged to share some of the new ways we have been incorporating technology in mathematics instruction through our methods class. There were many teachers who were not aware of the base ten virtual manipulatives website that had great interactive virtual manipulatives to teach addition with

> regrouping. In this way, I was able to bring to the table a new innovative teaching strategy and tool to enhance the lesson.

By working with special educators and instructors for English language learners, teachers discussed how to adapt tasks to meet students' Individualized Education Program goals, giving them access to meaningful mathematics. To generate ways to differentiate and scaffold instruction, teachers created a concept map outlining key components of prerequisite mathematics and interrelated concepts that might be future knowledge building blocks. A teacher with eight years of classroom experience reported:

> The mapping of prior knowledge needed and future knowledge was illuminating – it just got me thinking more deeply about the concept. The brainstorming helped to see what kids need to know and where they are headed. It makes it easy to see all of the standards that are tied into one concept. I learned about multiple models of representations and strategies.

The reciprocal learning allowed for everyone to build mathematical knowledge for teaching in terms of concepts, models, strategies and representations.

Revealing gaps in mathematical knowledge for teaching. Using an observational approach, researchers took anecdotal notes and collected in-depth information about teacher behaviors and any comments during the collaborative planning and debriefing sessions. To overcome researcher bias, the researchers used an observation checklist to document the level of input from all during collaborative planning (see Figure 3). The observation checklist included many of the practice-based skills identified as mathematical knowledge for teaching (Ball, 2003). Using tally marks, observers marked when preservice or practicing teachers made contributions to add to the mathematical knowledge for teaching.

	Practice based skills focused on during collaborative planning	Preservice teachers	Inservice teachers
1	Developing students' understanding of mathematics beyond algorithms.		
2	Taking students' prior understanding into account when planning curriculum and instruction.		
3	Engaging students in inquiry-oriented activities.		
4	Designing the instructional sequence that is appropriate and meaningful		
5	Assessing students' mathematical learning through questioning and take the next steps.		
6	Posing good mathematical questions and problems that are productive for students' learning.		
7	Making judgments about the mathematical quality of instructional materials and modify as necessary.		
8	Anticipating students' mathematical questions, curiosities, and misconceptions.		
9	Using mathematically appropriate and comprehensible explanations for students.		
10	Use technology with students.		
11	Giving access for mathematical learning to all members of a diverse population.		
12	Identifying and making connections among various mathematical topics.		
13	Representing mathematical ideas and concepts carefully in multiple ways.		
14	Making connections between physical, graphical models and symbolic notation.		
15	Generating novel teaching strategies		
	Use tally marks as contributions are made by preservice and inservice teachers		

Figure 3. Observation checklist used during lesson study sessions

In addition, the checklist served as a systematic way to discuss the planning process with the preservice teachers during a separate focus interview after the collaborative planning session. One researcher had the preservice teachers consider the 15-practice-based skills listed in the observation checklist and discuss what seemed most challenging or surprising. Although instructional practices and skills were discussed in the methods class, preservice teachers had not fully understood what practice-based skills were involved in the actual act of planning and teaching until the lesson study experience. They expressed that limited prior experiences with students made it challenging to anticipate students' mathematical questions, curiosities, and misconceptions and to take students' prior understanding into account when planning lessons. Other notable challenges were assessing and posing good mathematical questions and problems. As novice teachers, many had limited vertical mathematical knowledge and understanding of scope and sequence in mathematics. For example, in a conversation where teachers mapped prior and future mathematics learning, one preservice teacher asked such questions as, "When do they learn to divide by decimals? Should they be able to convert fractions to decimal by this grade?" The necessary depth of content knowledge and sequencing of mathematical ideas are generally learned as teachers gain experience in multiple grade levels or through vertical articulation across grade levels. Preservice teachers benefited from having the experience of collaborative planning with practicing teachers who helped them recognize practice-based skills using concrete examples.

Increasing awareness of the complexity of teaching and reflective practice. Through the collaborative reflection process, preservice teachers experienced the complexity of teaching firsthand. They experienced how carefully teachers select and set mathematical tasks; support student exploration of the task through questioning, use of representations and extensions; orchestrate a rich discussion to share ideas; and identify next steps to build upon student mathematical understanding. One preservice teacher commented on the sequencing and planning of activities:

> I was amazed that even experienced teachers wrestle with the ideas that we do when we plan lessons, like how to hook the students and link and engage the students. I thought it just came to them so naturally since they make it seem so easy when I observe them teach. Now I see how much thought is put into the actual sequencing of a lesson.

Another noted the attention paid to choosing appropriate mathematical tasks:

> It was really eye opening to see how the teachers had to pick and choose which mathematical model to use for the lesson and how to design the task sheet so that students could reveal their learning. There were even times when teachers questioned each other about the use of certain models fearing that it may confuse students down the road and whether using multiple models might actually confuse the special needs learners.

As they watched practicing teachers negotiate, problem solve, assist each other, and elaborate on each other's ideas, preservice teachers developed a vision of a community of practice with reflective practitioners. This experience, situated in an authentic teaching context, produced a level of understanding about lesson planning difficult to generate in a methods class. Preservice teachers wanted more opportunities to collaborate in such meaningful ways as reflecting on lessons, participating in joint problem solving, and collaborative planning.

Norms to Ensure Success in Lesson Study

During the lesson study process, a critical set of norms were established to ensure the success of the study for both preservice and inservice teachers:

- T rust and safety. The professional learning environment had to be free from and not linked to any form of evaluation of teachers or teaching for both preservice and practicing teachers. This safe environment allowed individuals to reveal insecurities and any fragile understanding of mathematical concepts, thus lessening anxiety. The safe environment raised teachers' productive disposition towards exploring their craft and analyzing student learning.

- Knowledge and competencies. Teachers recognized that different team members had different expertise and competencies, which developed a sense of collective efficacy; i.e., the team had a confident expectation that it would successfully achieve its intended goal.
- Shared experience and language. Their common experience helped teachers build collective knowledge as they worked together to understand and to make sense of challenges and appreciate the "aha" or surprise moment while analyzing student learning and seeing instructional improvement with new eyes.
- Lesson study facilitation. Designating a lesson study facilitator (a researcher) who could continue to engage teachers in studying the complex nature of teaching and learning mathematics, despite all the demands of teaching, helped sustain the learning enterprise.

Final Thoughts

Giving preservice teachers an opportunity to collaborate with practicing teachers at a school site supports Lave and Wegner's (1990) notion of situated learning: knowledge needs to be presented in authentic contexts, settings and situations normally involving that knowledge. Social interaction and collaboration with practicing teachers allowed preservice teachers to integrate classroom reality with the theory they learned in class. As a result, discussions during the lesson study sessions were qualitatively different than discussions typically found in the researchers' methods class. For example, preservice and experienced teachers alike struggled with effectively differentiating lessons for individual students. In the methods class, preservice teachers tended to discuss differentiation strategies with a general group of students (e.g., "English language learners" or "special needs learners"). However, during the lesson study experience, specific students with specific needs gave preservice teachers first-hand experience with how a lesson must meet those needs.

Additionally, preservice teachers' reflections indicate that anticipating student responses was the most challenging aspect of the four-column lesson plan. One preservice teacher stated, "I have never taught elementary students before so I am not sure what they will have challenges with in the following lesson." Another preservice teacher commented, "Hearing…teachers who taught this in previous years describing in detail what students had misconceptions about, and listening to how they talk about the common mistakes they make on assessments, helped me see how important assessment is to planning for instruction."

To help preservice teachers develop the mathematical knowledge needed for teaching, it is important that mathematics educators place them in situated learning contexts like this lesson study experience. Collaboration and reflection help preservice teachers develop professional dispositions as career educators who will continually reflect on practice and share learning with colleagues. This lesson study provided an opportunity for collaborative reflection where teachers openly shared instructional practices while developing relationships and an infrastructure for a continuous collaborative mentoring community.

References

Ball, D. (2003). *What mathematical knowledge is needed for teaching mathematics?* Paper presented at the February 6, 2003 Secretary's Summit on Mathematics, Washington, DC. Retrieved from http://www.erusd.k12.ca.us/ProjectALPHAweb/index_files/MP/BallMathSummitFeb03.pdf

Chazan, D., Ben-Chaim, D., Gormas, J., Schnepp, M., Lehman, M., Bethell, S. C., & Neurither, S. (1998). Shared teaching assignments in the service of mathematics reform: Situated professional development. *Teaching and Teacher Education, 14*, 687–702.

Fernandez, C., & Yoshida, M. (2004). *Lesson study: A case of a Japanese approach to improving instruction through school-based teacher development.* Mahwah, NJ: Lawrence Erlbaum.

Hawbaker, B. W., Balong, M., Buckwalter, S., & Runyon, S. (2001). Building a strong BASE of support for all students through co-planning. *Teaching Exceptional Children, 33*(4), 24–30.

Kolb, D. (1984). *Experiential learning as the science of learning and development.* Englewood Cliffs, New Jersey: Prentice Hall.

Lave, J., & Wenger, E. (1990). *Situated learning: Legitimate peripheral participation.* Cambridge, UK: Cambridge University Press.

Lewis, C. (2002). *Lesson study: A handbook of teacher-led instructional change* Philadelphia, PA: Research for Better Schools.

Lewis, C., Perry, R., & Murata, A. (2006). How should research contribute to instructional improvement? The case of lesson study. *Educational Researcher, 35*, 3–14.

Ma, L. (1999). *Knowing and teaching elementary mathematics: Teachers' understanding of fundamental mathematics in China and the United States.* Mahwah, NJ: Lawrence Erlbaum.

Roth, W. M., & Tobin, K. (2004). Coteaching: From praxis to theory. *Teachers and Teaching: Theory and Practice, 10*, 161–180.

Smith, M. S., Bill, V., & Hughes, E. K. (2008). Thinking through a lesson protocol: Successfully implementing high-level tasks. *Mathematics Teaching in the Middle School, 14*, 132–138.

Strauss, A., & Corbin, J. (1994). Grounded theory methodology: An overview. In N. K. Denzin & Y. S. Lincoln (Eds.), *Handbook of qualitative research* (pp. 1–18). London: Sage Publications.

Walther-Thomas, C. S. (1997). Co-teaching experiences: The benefits and problems that teachers and principals report over time. *Journal of Learning Disabilities, 30*, 395–407.

Dr. Jennifer Suh is an assistant professor of mathematics education at George Mason University in Fairfax, Virginia. Her research focuses on developing teachers' pedagogical content knowledge in mathematics through lesson study and building students' mathematical proficiency through problem solving. She can be reached at jsuh4@gmu.edu

Jana Parker is a doctoral candidate at George Mason University and currently an instructional coach at DC Public Schools. Her research focuses on how co-teachers' collaboration and joint professional learning facilitate the implementation of standard-based mathematics. She can be reached at jparker9@gmu.edu

Thompson, D. R.
AMTE Monograph 7
Mathematics Teaching: Putting Research into Practice at All Levels
© 2010, pp. 141–155

10

Reading in the Content Area: A Mathematics-Specific Course Design

Denisse R. Thompson
University of South Florida

Many states require teachers to complete professional development focused on content literacy. Yet such work is often generic in nature so that mathematics teachers must determine how to apply the professional development to their own mathematics classroom. This paper describes a mathematics-specific content literacy course required in a preservice teacher preparation program. Course objectives and the weekly focus of the course are discussed in detail. In addition, sample assessments and student perspectives are shared. Sufficient information and resources are included so that other mathematics teacher educators could build such a course at their own institutions or develop a major unit on content literacy for inclusion in a methods course.

Since at least the advent of the *Curriculum and Evaluation Standards for School Mathematics* (National Council of Teachers of Mathematics [NCTM], 1989), it has been recommended that communication be part of mathematics instruction. Mathematics teachers are encouraged to develop a language-rich classroom in which their students use literacy to help make sense of mathematics. Many initial certification programs require preservice teachers to complete a "reading in the content area" course. In a survey of state departments of education, Romine, McKenna, and Robinson (1996) found that 47 states and the District of Columbia required secondary content teachers to have a specific course or specified competencies related to content literacy. However, they also found that such courses needed to support implementation of effective literacy practices or little change was likely to occur in actual classrooms.

In Florida, although secondary content teachers have been encouraged to complete professional development focused on content literacy, mathematics teachers may be left to "fend for themselves" in applying literacy strategies to mathematics. This paper outlines a mathematics-specific content reading course at a Florida university that addresses this need. It includes an outline of major topics, assignments, and rubrics used to help preservice teachers (students) engage with mathematical literacy, as well as observations and reactions regarding the value of various assignments. This 3-credit course, required of all preservice secondary teachers receiving mathematics certification in grades 6–12, meets once per week throughout the semester with students typically completing the course within two semesters of internship.

Overview of the Course

The course adopts a broad view of literacy including reading, writing, listening, and speaking in building a language-rich environment. The purpose of the course is to provide students an opportunity to develop or expand concepts, skills, and instructional procedures for effectively integrating communication in its broad view into the mathematics curriculum. Objectives for the course promote knowledge and application of the following:

- Relationships among language development, reading, and the teaching and learning of mathematics.
- Characteristics of the language of mathematics (vocabulary and symbolism) and the major problems students encounter with this language.
- Instructional strategies to help students improve skills in reading mathematics and developing mathematical vocabulary.

- Problem-solving processes and instructional procedures to aid in the solution of verbal mathematical problems.
- Ways to integrate writing into the mathematics classroom and procedures for helping students improve their writing skills in mathematics.
- Techniques for assessing and evaluating open-ended assignments, including the development and application of rubrics.

The course blends theoretical issues related to mathematics language with practical strategies appropriate for classroom implementation. Most prospective teachers in the course initially believe that literacy is an "extra" activity that takes time away from teaching content. By integrating theory with practice, they realize that mathematical literacy can be implemented as a normal part of instruction. In fact, literacy strategies can help them understand their future students' thinking and appropriately modify instruction so students will be mathematically successful.

The following sections outline the major topics discussed in the reading course each week. A significant part of the course focuses on technical aspects of mathematics language and issues related to reading, writing, and discourse, along with strategies for addressing these. The course also includes exploration of appropriate assessments that help form the basis of grades.

Weeks 1–2: A Rationale for Mathematical Literacy and Discourse

The first week sets the stage for the importance of literacy within the broader scope of mathematics education. In the first class session, activities are introduced that might be used at the beginning of a school year to orient future students of the prospective teachers to a classroom where mathematical literacy has an integral role. For instance, after reading *Math Curse* (Scieszka & Smith, 1995), prospective teachers are shown a variety of high school mathematics tasks related to the book. As a result, they begin to recognize engagement with literacy in formats that provide opportunities to review numerous topics (e.g., measurement, rational numbers). Students in the course also reflect on their own experiences with literacy in mathematics classes. They usually recognize a need to read mathematics textbooks, but often struggle with doing so. Some mention specific classes that required significant writing. Expectations of embedding literacy in mathematics instruction are not entirely unfamiliar to them, even if rare.

Because mathematical literacy often seems foreign to students' experiences, discussion focuses on national recommendations regarding the importance of mathematical literacy. In particular, the course references the Communication Standard in the *Principles and Standards for School Mathematics* (NCTM, 2000) and Teaching Standard 2: Discourse in the *Professional Standards for Teaching Mathematics* (NCTM, 1991). Both standards highlight teachers listening to oral questions and responses, expecting justification of answers, helping students learn to use mathematical language to express ideas, and determining when to introduce formal mathematical language and notation.

During the second week, prospective teachers focus on discourse in the classroom. Using work developed by Goldin (1998) and adapted with colleagues (Beckmann, Thompson, & Rubenstein, 2010), they think about the nature of questions they will ask their future students. Goldin developed a progression of question types that provide "just enough" help for students working through problems. His work, the basis for a Stages of Questioning framework, allows teachers to help students without removing the cognitive demand of the task (Beckmann, et al., 2010). For instance, teachers start by using non-directive questions to help students identify a problem, then suggest general problem-solving strategies such as drawing a picture, followed by other strategies specific to the given task (e.g., identifying a particular pattern or rule). Finally, teachers help learners think about their own thinking (e.g., metacognition). Through this discussion of the Stages of Questioning framework and of research by Hufferd-Ackles, Fuson, and Sherin (2004) on levels of teacher-student interaction, students in the course consider how different types of questions promote or stifle discourse and communication.

Weeks 3–8: The Language of Mathematics and Communication Strategies

To encourage literacy in future classrooms, prospective teachers need to understand issues of mathematical language and strategies incorporated in the classroom that might help with those issues. Numerous authors have highlighted difficulties existing with mathematics language (Adams, 2003; Gay, 2008; Krussel, 1998; Miller, 1993; Rubenstein & Thompson, 2001; Thompson & Rubenstein, 2000; Usiskin, 1996). This portion of the course is designed to identify difficulties with language that must be confronted and addressed. Some difficulties of mathematics language include the following:

- Words may have different meanings in mathematics and in English (e.g., product, plane).
- Words may have more than one mathematical meaning (e.g., square, base, median).
- Words may only have a mathematical meaning (e.g., hypotenuse).
- Phrases need to be learned in their entirety (e.g., if-and-only-if).
- Mathematics symbols may require multiple words to verbalize (e.g., $\leq$).
- Some concepts are expressed via multiple symbols (e.g., multiplication or division). (Thompson et al., 2008)

Once students in the course have developed a background in issues of language, strategies to incorporate reading and writing into the mathematics classroom are discussed: concept worksheets (Toumasis, 1995); link sheets (Shield & Swinson, 1996); reading quizzes (Rubenstein, 1992); prereading, during reading, and post-reading strategies (Daniels & Zemelman, 2004); RAFT writing tasks (Barton & Heidema, 2002); and word roots (McIntosh, 1994; Rubenstein, 2000). Through these discussions students become conversant with many strategies and begin to acquire a variety of options in their literacy instruction repertoire. They determine which ones resonate with their own teaching perspectives and have the best potential for use in their classrooms and in yearlong planning.

Because secondary students build understanding as they use mathematics language to process ideas, their teachers must be deliberate in the use and discussion of mathematics language. Murray (2004) suggests that students need to use a word at least 30 times in order to own it. Making this type of connection is one purpose of the discussions throughout this portion of the course.

Week 9: Children's Literature

Although many educators use children's literature in teaching elementary mathematics, few have written about its use at the middle or high school level (Austin, Thompson, & Beckmann, 2003; Bay-Williams & Martinie, 2004; Beckmann, Thompson, & Austin, 2004). This course introduces a range of children's literature applicable to the middle and/or high school level. For example, *The King's Chessboard* (Birch, 1988) provides an opportunity to explore exponential functions, while *Gulliver's Adventures in Lilliput* (Beneduce, 1993) explores the use of similarity and ratio within the context of a classic tale.

Weeks 10–11: Word Problems

To support in-depth work with word problems, applications from first- and second-year algebra are solved from multiple perspectives (symbolic, graphical, tabular, and pictorial). Consider the following samples:

- Alaska's population of 480,000 has been increasing at a rate of 6000 people a year. Delaware's population of 606,000 has been increasing at a rate of 4000 people a year. If these rates of increase continue, in how many years will the populations be equal? (adapted from McConnell, et al., 1990, p. 287)
- In 1987, the world population passed the 5 billion mark. If the population is growing at an annual rate of 1.6%, in what year will the population exceed 15 billion? (adapted from Senk, et al., 1990, p. 512)

Class discussion of these problems centers on difficult aspects for secondary students as well as different solution approaches. Reading strategies applied to the problems, along with Pólya's (1957) problem-solving process, encourage students to break down barriers by considering alternative approaches to problems. For instance, one strategy discussed is using tables and patterns as seen in Table 1 to build expressions for the Alaska-Delaware problem.

Table 1
Using Tables and Patterns to Build Expressions in Word Problems

Year	Alaska Population	Alaska Population Written Another Way
0	480,000	480,000
1	480,000 + 6000	480,000 + 6000
2	480,000 + 6000 + 6000	480,000 + 2 • 6000
3	480,000 + 6000 + 6000 + 6000	480,000 + 3 • 6000
...	...	...
n		480,000 + n • 6000

Examining the value of leaving computations unfinished in order to expose patterns leads to writing symbolic representations for the problem. Once equations are developed, graphs, tables, or symbolic manipulators are used to find an actual solution.

Weeks 12–14: Assessment

In the final portion of the course, the integration of literacy with formative and summative assessment is discussed. If teachers require their students to write about mathematics as part of instruction, then consideration must be given to how such writing is assessed. As one example, appropriate tasks for in-class tests requiring students to explain and justify their thinking demand the use of and practice with scoring rubrics.

Course Assessments

Three major assignments are used to assess the mathematics content reading course: (a) a reading assignment, i.e., a book report on a book about mathematics (15% of course grade); (b) a collection of children's literature resources and related tasks (20% of course grade); and (c) a year-long unit plan to integrate literacy into the mathematics classroom (50% of course grade). The remainder is based on class participation.

Reading Assignment

The reading assignment, given on the first day of class, provides a learning experience before the prospective teachers have enough background to do a substantive assessment related to literacy. Each student in the course reads a non-text mathematics book, writes a 3- to 5-page paper, and shares an abstract of the book with peers. The paper must include the following information:

1. A complete bibliographic citation
2. A brief summary of the book highlighting essential features
3. An indication of an appropriate school audience with a brief rationale citing the book's strengths or weaknesses
4. An indication of how this book might be used with a mathematics class
5. The reviewer's personal reaction to the book
6. A one-page abstract containing the citation, an abbreviated summary, the intended audience, and an abbreviated reaction to the book

This assessment gives students some ideas about mathematical reading, and seems to be a good beginning for an emphasis on mathematical literacy. Consider the following comments:

- On reading *Why do buses come in threes?* (Eastaway & Wyndham, 1998): "It actually helped me understand math formulas and the reasoning behind these formulas that I have never been taught, so not only was this book a good tool that could be used for middle or high school students, but it was also a great tool for me as a college student."

- On reading *Go figure!* (Ball, 2005): "I believe this book is appropriate for any classroom setting because it can help a student who has lost motivation in a mathematics class.... It shows how math is a part of other subjects.... It proved to be an easy read with less than 100 pages, but I feel I learned more about math from this 100-page book than I would ever learn in a 500-page textbook."

Students are often quite surprised at how much they enjoy and learn from the assignment. One wrote, "The world of math literature was opened up to me for the first time. I benefited by being introduced to this wealth of information, as well as being required to write about how I would incorporate it into my future classes." Another said, "Just finding a book really helped. I never knew there were so many books about math."

Literature Assignment

Week 9's literature assignment engages students by showing them many children's books exist that can be used in mathematics. A shortened version of the assignment is presented below:

1. Find 15 books, short stories, poems, and/or videos appropriate for use with middle school or high school students. For each, write two questions or brief activities for use in class. Questions should elicit more than just a right/wrong answer, should have a variety of possible answers, and should require some higher-order thinking skills.
2. For each resource, provide a brief summary, the bibliographic citation, and the questions/activities. Provide sample responses.
3. Create a matrix indicating the content strand (Number and Operations, Measurement, Geometry and Spatial

Sense, Algebraic Thinking, and Data Analysis and Probability) that the questions and activities address.

With this assignment, students find resources and write questions of varying levels of difficulty, as indicated in the following sample that references *The Treasure* (Shulevitz, 1978).

- Isaac has walked many miles to arrive at the king's castle…. Find out some facts about the Appalachian National Scenic Trail. Let's assume that Isaac's house was at one end of the trail and the castle at the other end. If Isaac was walking 3 miles/hour, stopping to eat and sleep for 8 hours a day, how many days did his roundtrip last?
- According to the U.S. Department of Health and Human Services, it is estimated that close to 40 million Americans live below the poverty line. Analyze the table and answer the following questions. [A data table was included with questions requiring interpretation.]

Students sometimes complain about the number of books they are required to find and/or the difficulty of writing good questions. Nevertheless, they typically are surprised at their own creativity and many perceive the benefits of this literacy approach. One prospective teacher commented, "I was introduced to another facet of mathematical literature…. I fell in love with this way of teaching and introducing new concepts through the written word!" Another noted, "It helped me get an idea of how I can incorporate reading in my classroom or perhaps how I can introduce a concept." Those who are engaged in field experiences often try some of the books with middle/high school students and are surprised at the positive reactions of the students.

Communication Unit Plan

This major assignment is designed to have prospective teachers think about how they might embed literacy in their future classrooms throughout the year. The students may work together or alone, but must choose a grade level or course on which to focus. They are limited to no more than ten chapters of a text or ten major course topics. For each chapter, they must address the multiple views of literacy mentioned earlier in this article and develop five real-life word problems and two alternative assessment items for student communication of mathematical understanding; they also create two unit tests correlated to objectives and reflective of important communication in the classroom.

Built-in checkpoints help ensure that students are on the right track and do not procrastinate in working on the assignment, and guarantee that the work will meet the instructor's expectations. The goal is for students to determine how to make literacy an integral part of instruction and to demonstrate it in this assignment. Typically a notebook-sized portfolio of literacy activities is the result of each student's effort. The following comments reflect positive reactions to this assignment:

- "Being forced to put 'feet' to all of the theory and good ideas that we have been learning was of benefit to me."
- "This was a great way of bringing all the course material together. It shows me how the techniques and ideas we learned can be applied to teaching a real math course."
- "It was beneficial to use all the different vocabulary, reading, writing, and alternative assement [sic] tools, that we learned in class, in a year-long plan....I'm sure [it] has made us much better prepared for our teaching careers."

Conclusion

Mathematics teachers need strategies for building literacy in their classrooms and specifically for using literacy with mathematics content. A mathematics-specific content reading course may provide a way to address deficiencies in generic courses. This paper provides a model for the development of such a course. It provides the basis for a unit on content literacy within a mathematics methods course and suggestions for generic content literacy instructors to make specific connections to mathematics.

Student comments on class assignments suggest that students' views about the nature of literacy can be positively impacted by such a course. Mathematics teacher educators are preparing their students for classrooms in which there are expectations for reading and writing mathematics and for communicating about solving problems. To incorporate literacy effectively into their future classrooms, prospective teachers need experiences that challenge their views of mathematics instruction and that provide specific examples about what mathematical literacy might look like in action. The model proposed here, in full or in part, provides a base upon which mathematics teacher educators can build substantial experiences with mathematical literacy for prospective teachers.

References

Adams, T. L. (2003). Reading mathematics: More than words can say. *The Reading Teacher, 56*, 786–795.

Austin, R. A., Thompson, D. R., & Beckmann, C. E. (2003). Exploring measurement concepts through literature: Natural links across disciplines. In D. H. Clements & G. Bright (Eds.), *Learning and teaching measurement* (pp. 245–255). Reston, VA: National Council of Teachers of Mathematics.

Ball, J. (2005). *Go figure! A totally cool book about numbers*. New York: DK Publishing, Inc.

Barton, M. L., & Heidema, C. (2002). *Teaching reading in mathematics: A supplement to* teaching *reading in the content areas: If not me, then who?* (2nd ed.). Alexandria, VA: Association for Supervision and Curriculum Development.

Bay-Williams, J. M., & Martinie, S. L. (2004). *Math and literature: Grades 6–8*. Sausalito, CA: Math Solutions.

Beckmann, C. E., Thompson, D. R., & Austin, R. A. (2004). Exploring proportional reasoning through movies and literature. *Mathematics Teaching in the Middle School, 9,* 256–262.

Beckmann, C. E., Thompson, D. R., & Rubenstein, R. N. (2010). *Teaching and learning high school mathematics*. Hoboken, NJ: John Wiley & Sons, Inc.

Beneduce, A. K. (1993). *Gulliver's adventures in Lilliput*. New York: Putnam & Grosset.

Birch, D. (1988). *The king's chessboard.* New York: Puffin Pied Piper Books.

Daniels, H., & Zemelman, S. (2004). *Subjects matter: Every teacher's guide to content-area reading*. Portsmouth, NH: Heinemann.

Eastaway, R., & Wyndham, J. (1998). *Why do buses come in threes? The hidden mathematics of everyday life*. New York: John Wiley & Sons.

Gay, A. S. (2008). Helping teachers connect vocabulary and conceptual understanding. *Mathematics Teacher*, *102*, 218–223.

Goldin, G. A. (1998). Observing mathematical problem solving through task-based interviews. In A. R. Teppo (Ed.), *Qualitative research methods in mathematics education* (pp. 40–62). Reston, VA: National Council of Teachers of Mathematics.

Hufferd-Ackles, K., Fuson, K. C., & Sherin, M. G. (2004). Describing levels and components of a math-talk learning community. *Journal for Research in Mathematics Education*, *35*, 81–116.

Krussel, L. (1998). Teaching the language of mathematics. *Mathematics Teacher*, *91*, 436–441.

McConnell, J. W., Brown, S., Eddins, S., Hackworth, M., Sachs, L., Woodward, E., Flanders, J., Hirschhorn, D., Hynes, C., Polonsky, L., & Usiskin, Z. (1990). *The University of Chicago School Mathematics Project algebra*. Glenview, IL: Scott, Foresman and Company.

McIntosh, M. E. (1994). Word roots in geometry. *Mathematics Teacher*, *87*, 510–515.

Miller, L. D. (1993). Making the connection with language. *Arithmetic Teacher*, *40*, 311–316.

Murray, M. (2004). *Teaching mathematics vocabulary in context*. Portsmouth, NH: Heinemann.

National Council of Teachers of Mathematics. (1989). *Curriculum and evaluation standards for school mathematics*. Reston, VA: Author.

National Council of Teachers of Mathematics. (1991). *Professional standards for teaching mathematics*. Reston, VA: Author.

National Council of Teachers of Mathematics. (2000). *Principles and standards for school mathematics*. Reston, VA: Author.

Pólya, G. (1957). *How to solve it: A new aspect of mathematical method* (2nd ed.). Princeton, NJ: Princeton University Press.

Romine, B. G. C., McKenna, M. C., & Robinson, R. D. (1996). Reading coursework requirements for middle and high school content area teachers: A U. S. survey. *Journal of Adolescent & Adult Literacy, 40*, 194–198.

Rubenstein, R. N. (1992). Tips for beginners: Improving students' reading with quizzes. *Mathematics Teacher, 85*, 624–635.

Rubenstein, R. N. (2000). Word origins: Building communication connections. *Mathematics Teaching in the Middle School, 5*, 493–498.

Rubenstein, R. N., & Thompson, D. R. (2001). Discuss with your colleagues: Learning mathematical symbolism: Challenges and instructional strategies. *Mathematics Teacher, 94*, 265–271.

Scieszka, J., & Smith, L. (1995). *Math curse*. New York: Viking.

Senk, S. L., Thompson, D. R., Viktora, S. S., Rubenstein, R., Halvorson, J., Flanders, J., Jakucyn, N., Pillsbury, G., & Usiskin, Z. (1990). *The University of Chicago School Mathematics Project advanced algebra*. Glenview, IL: Scott, Foresman and Company.

Shield, M., & Swinson, K. (1996). The link sheet: A communication aid for clarifying and developing mathematical ideas and processes. In P. C. Elliott and M. J. Kenney (Eds.), *Communication in mathematics, K–12 and beyond* (pp. 35–39). Reston, VA: National Council of Teachers of Mathematics.

Shulevitz, U. (1978). *The treasure*. Toronto: McGraw-Hill.
Thompson, D. R., Kersaint, G., Richards, J. C., Hunsader, P. D., & Rubenstein, R. N. (2008). *Mathematical literacy: Helping students make meaning in the middle grades*. Portsmouth, NH: Heinemann
Thompson, D. R., & Rubenstein, R. N. (2000). Learning mathematics vocabulary: Potential pitfalls and instructional strategies. *Mathematics Teacher*, *93*, 568–574.
Toumasis, C. (1995). Concept worksheet: An important tool for learning. *Mathematics Teacher*, *88*, 98–100.
Usiskin, Z. (1996). Mathematics as a language. In P. C. Elliott and M. J. Kenney (Eds.), *Communication in mathematics, K–12 and beyond* (pp. 231–243). Reston, VA: National Council of Teachers of Mathematics.

Denisse R. Thompson is professor of mathematics education at the University of South Florida, Tampa, FL, where she teaches a range of courses for K–12 teachers. Her interests include curriculum development/evaluation, assessment, language/literacy, and bringing culture into the mathematics classroom. She can be reached at denisse@usf.edu.

Van Zoest, L. R., Stckero, S. L., and Edson, A. J.
AMTE Monograph 7
Mathematics Teaching: Putting Research into Practice at All Levels
© 2010, pp. 157–171

11

Multiple Uses of Research in a Mathematics Methods Course

Laura R. Van Zoest
Western Michigan University

Shari L. Stockero
Michigan Technological University

Alden J. Edson
Western Michigan University

This chapter focuses on putting research into practice in the design and implementation of a mathematics methods course. Illustrations are provided of using others' research to inform practice, supporting prospective teachers in developing habits reminiscent of the research process, and engaging in research to systematically answer questions about the course itself. Putting research into practice in these multiple ways supports both prospective teachers and their teacher educators in improving their respective classroom practices. Prospective teachers develop tools, knowledge and dispositions needed to ground reflections on practice in classroom data and use knowledge from the broader mathematics education community to inform instruction. Teacher educators use data from the methods course in conjunction with research literature to inform their own practice.

Holding dual roles of researcher and teacher educator and choosing to ground their research in methods courses for future teachers have provided the authors with opportunities to put research into practice. Specifically, they incorporated the following aspects of research into the content and design of a secondary mathematics methods course: results from other researchers, elements of the research process, and results from their own research. This paper elaborates on these aspects to illustrate a variety of ways research can be put into practice in a methods course, and, by extension, into other areas of mathematics teacher educators' practice.

Using Others' Research Results

The work of mathematics education and teacher education researchers has significantly influenced the authors' decisions in designing the methods course described here, planning course activities, implementing the planned activities, and providing feedback to prospective teachers on their learning. For example, the course design was informed by Simon and Tzur's (2004) work on learning trajectories, Darling-Hammond and Hammerness' (2005) discussion of the importance of coherence and connectedness within a course, Ball and Cohen's (1999) arguments for learning from practice in practice, and Ball and colleagues' work on mathematical knowledge for teaching (see, e.g., Ball, Thames, & Phelps, 2008).

The work of other researchers has also provided content for the course. In particular, written reports of others' research inform prospective teachers' thinking and support work in meeting course goals (see Van Zoest & Stockero, 2008c). Course readings provide prospective teachers with details about mathematics teaching and learning to which they may not otherwise have access, as well as a common language for engaging in conversations about course ideas. For example, research originating with the QUASAR project (Silver & Stein, 1996) provides rich frameworks for talking about the relationship between tasks, task implementation, and student learning. Another research example affecting both the content and form of the course is work on complex instruction (Cohen, Lotan, Scarloss, & Arellano, 1999). In the following sections, these dissimilar bodies of research are used to illustrate multiple ways in which the course incorporates others' research results.

Mathematics Tasks Framework

The Mathematics Tasks Framework (Stein & Smith, 1998) and the Task Analysis Guide (Stein, Smith, Henningsen, & Silver, 2000) guided the selection of high-level tasks for the methods course. These tasks allow multiple solution strategies and representations while not providing a suggested way to approach the task. In addition, Stein and Smith's (1998) related Factors Associated with the Maintenance and Decline of High-Level Cognitive Demands aided in maximizing the tasks by focusing attention on issues related to task implementation. For example, a focus on encouraging prospective teachers to provide justification beyond algorithmic knowledge challenges them to reconsider often-fragile mathematical understandings and to make connections among mathematical ideas.

Prospective teachers intuitively know that course tasks are different than those typically used in their own school mathematics learning; the Mathematics Tasks Framework provides a tool for discussing what makes the course tasks and their implementation different. Prospective teachers first engage in a task-sort activity (Smith, Stein, Arbaugh, Brown, & Mossgrove, 2004) classifying a number of secondary-grades-mathematics tasks as either high- or low-level cognitive demand. Ensuing discussion around the thinking involved in sorting the tasks provides an opportunity to introduce the Task Analysis Guide and make explicit the characteristics of tasks used in the course that make them high-level. Prospective teachers then read Stein and Smith's (1998) article, which includes the Factors Associated with the Maintenance and Decline of High-Level Cognitive Demands. These frameworks become tools for analyzing teaching and learning, as well as mathematics curricula, throughout the course. For example, to clarify Stein and Smith's (1998) ideas and develop skills in using these research results, prospective teachers use the frameworks to analyze other teachers' practice in video clips. They also use them in written assignments focused on analyzing their own practice in the course field experience. Rather than simply identifying parts of a lesson that did or did not go well and speculating why, the frameworks provide a concrete way to identify factors affecting task implementation and to consider alternate teacher moves that may have better supported students' task work.

Complex Instruction

Complex instruction is a collection of strategies demonstrated to increase learning opportunities for students (Cohen et al., 1999). At its core is the use of challenging small group tasks that include multiple ways for students of varying abilities to contribute to the groups' work. One strategy, assigning students to roles when working on group tasks, creates an environment where each individual's success depends on supporting the learning of the group. Another important strategy, assigning competence, allows the teacher to mediate status issues stemming from the value students place on each other's perceived intellectual, social, and other abilities, by publicly recognizing contributions of low-status students.

Complex instruction is incorporated into the methods course to provide prospective teachers with tools for supporting learners in an inclusive classroom setting where differentiated instruction is necessary to meet all students' needs. In addition to providing prospective teachers with an academic knowledge of complex instruction, the principles of complex instruction were applied to instruction in the methods course. In particular, modeling complex instruction strategies by implementing group roles and assigning competence among prospective teachers in challenging contexts made their learning experiential as they saw firsthand how these techniques affect group dynamics and change group member status. Explaining the reasons for engaging in these activities allowed prospective teachers to see how their instructors' practices incorporated the research advocated for prospective teachers to use in their future practice in schools.

Not only is complex instruction useful for the future teaching of prospective teachers, it helped them to make sense of and address issues arising in the course field experience where they work with small groups of academically, racially, and economically diverse middle school students. The required reading on complex instruction (Cohen, 1998) and experiences with ideas in the course provided prospective teachers with a lens and language for analyzing interactions of middle school student groups as well as their own roles in facilitating those group interactions. Thus, incorporating this research into the methods course design allowed prospective teachers to engage with complex instruction in three ways: learning about complex instruction as an instructional tool; experiencing elements of complex instruction as learners; and applying complex instruction strategies and concepts to their own teaching.

Using Elements of the Research Process

To inform their practice, prospective teachers in the secondary mathematics methods course engage in activities reminiscent of conducting research in at least three ways: grounding ideas in literature; collecting and analyzing data; and drawing conclusions supported by evidence. Although the three are mutually supportive and synergistic, they are discussed separately in the following sections to highlight the contributions each makes to prospective teachers' learning.

Grounding Ideas in Literature

The critical role that readings drawn from the mathematics education literature play in supporting prospective teachers' learning has been described by Van Zoest and Stockero (2008b). Attention is paid to prospective teachers' use of course readings as a tool for analyzing and talking about teaching and learning mathematics. Most importantly, rather than as a launching point for discussion, they are used as a means to ground course discussions. Although conversations occasionally focus on the articles themselves, the primary purpose is to enable prospective teachers to incorporate the language and ideas from readings into discussions directly connected to classroom practice, including the use of such practice-based materials as mathematics tasks, lesson plans, video clips of instruction, and student work. For example, the concepts of funneling and focusing (Wood, 1998) are clarified through discussing examples and non-examples because they have proven useful to support prospective teachers' thinking about student questioning. It is then common for prospective teachers to discuss their own fieldwork-based questioning of students in terms of funneling and focusing, providing further opportunity to refine their understanding of the concepts.

Another illustration comes from a discussion regarding whether a student response from a field experience counted as justification. Rather than simply debating classmates' differing views, the article "Why, Why Should I Justify?" (Lannin, Barker, & Townsend, 2006) was used as a reference point for analyzing whether the response met the qualifications for justification described in the article. Using a reading to ground such discussions supports course goals by emphasizing the idea of mathematics education as a profession as it shifts the focus from personal preferences to drawing from the broad mathematics education community knowledge base, by providing prospective teachers opportunities to make connections between theory and practice, and by supporting prospective teachers in talking about practice in a disciplined way.

Incorporating literature into statements made about teaching and learning of mathematics is an expectation for class discussions and written work. Prospective teachers are not expected to parrot back a particular view about teaching and learning, but rather to compare their ideas with those they read and to analyze teaching and learning instances from the perspective of the readings. Prospective teachers are supported in meeting this expectation through explicitly designed pedagogical tasks that take advantage of existing interpretive frameworks in the literature, such as the Mathematics Tasks Framework and complex instruction research mentioned above, to guide analysis of teaching and learning. At the beginning of the course, the instructors establish a norm for class discussions to incorporate literature references by explicitly setting this expectation. This expectation is then reinforced by prompting for literature connections during discussions (e.g., "Does that remind you of anything you've read?") and providing positive reinforcement when prospective teachers make connections and use the ideas on their own. This practice sets the stage for similarly supporting written statements with references to the literature, an expectation reinforced by providing rubrics in advance that make it clear that incorporating connections to the literature is a critical component of assignments.

Collecting and Analyzing Data

Prospective teachers analyze classroom data, transcripts, student work, and collections of mathematics tasks throughout the course, and collect and analyze data from their own teaching during the related field experience (see Van Zoest & Stockero, 2009). After each audiotaped field experience teaching episode, the prospective teachers transcribe a portion of the session and use that in conjunction with students' written work to describe the students' mathematical understanding and ways in which teacher actions helped or hindered students' learning. Having concrete data to analyze has greatly improved prospective teachers' abilities to reflect on and learn from their experiences, both individually and collectively. For example, one prospective teacher's group of five students presented four different approaches to solving a mathematical task at the core of the teaching episode. His data were used to prompt a discussion about selecting and sequencing student solutions (Smith, Hughes, Engle, & Stein, 2009), allowing him to reflect on his responses to student ideas during this teaching episode and supporting the class's ability to orchestrate a discussion of student ideas. This iterative process of collecting, analyzing, and learning from classroom data is one that teachers can use to improve practice throughout their careers.

Drawing Conclusions Supported by Evidence

Central to mathematics education research is drawing conclusions based on data and, when sharing those conclusions with others, supporting them with evidence. Similarly, prospective teachers are expected to draw conclusions based on records of practice and to cite evidence when making their views public, either during class discussions or in written assignments. Initially the methods class instructors scaffold the development of this skill by encouraging students to support claims about teaching and learning seen in video clips of classroom instruction with line numbers from accompanying transcripts. Extensive work with practice-based materials in the course provides additional opportunities to reinforce and strengthen understanding and skills in this area (for more details, see Van Zoest & Stockero, 2008b).

Complex instruction, discussed earlier, was incorporated into the course primarily because of an observation that prospective teachers previously drew conclusions from their teaching experiences based on perceptions rather than evidence. For example, when a school student did not participate in one of the assigned field experience groups, prospective teachers quickly assumed the student did not care, was not putting forth enough effort, or was not capable of doing the work. From the methods course instructors' perspective, other evidence-based explanations might have better supported the prospective teachers' decision making in subsequent field experiences.

Collecting and analyzing data from field experience teaching episodes with such tools as the Mathematics Tasks Framework and complex instruction enabled the prospective teachers to notice a variety of issues, such as: posing a task that was too vague for the student to begin; a student's expressed fear of failure and the need for a safe environment in which to take risks; and the interesting unshared work of a student others considered "dumb." When encouraged to support conjectures about the issues with evidence, prospective teachers often could not find any for their initial conclusions. By uncovering evidence, however, they were led to surprising, yet empowering, conclusions pointing to ways that they, as teachers, could take actions to improve student learning. Overall, incorporating elements of the processes used by researchers into the methods course has given prospective teachers tools to support their work, broadened their ideas about mathematical thinking and learning, and helped them to develop a view of teaching as a professional activity.

Researching the Course

Like most teacher educators, the authors regularly use student feedback, written work, and assessment data to evaluate the learning that takes place in the methods course they teach. Over the past several years, the methods course has also served as a research site to address questions ranging from the effectiveness of specific course activities to the long-term effects of course learning outcomes. To support this research, data including video recordings of class sessions and copies of prospective teachers' written work are regularly and systematically collected. Although students are invited to participate in the research at the beginning of the class, in order to protect them from any potential negative effects of choosing not to participate, the instructors do not know whether individual students have agreed to participate until after course grades have been submitted.

The research team is a fluid group including both faculty teaching the course and faculty and graduate students not teaching the course. The involved instructors have found that opening the classroom to others provides multiple perspectives on teaching and learning in the course that would not be available if they individually reflected on their practice. The following examples demonstrate types of research conducted in the context of the course and describe some of the ways in which the results of this research have influenced the course design and instruction.

Example 1: Studying a Specific Component of the Course

A study of the Mathematics Teaching Autobiography required in the methods course addressed the broader question of how to most effectively scaffold prospective teacher learning and reflection. As described in Van Zoest and Stockero (2008a), this study was conceptualized based on informal observations of the traditionally assigned Mathematics Teaching Autobiography, which resulted in many students including only superficial statements about experiences and perspectives rather than thinking deeply about how prior experiences may have affected their views of teaching and learning. In an effort to understand and support the prospective teachers' analysis of the relationships among prior experiences, current views, and future teaching, the instructors introduced six synergistic scaffolds, including readings, a rubric, and targeted feedback, during the writing and revision process for the autobiography. Effects of using the scaffolds and prospective teachers' perspectives on their usefulness were then systematically analyzed with extensive coding and other research methods.

Although the instructor/researchers intuitively felt that the scaffolds introduced would support the prospective teachers' learning, the research helped identify which specific scaffolds were most supportive and in what ways. One surprising finding regarding the Mathematics Teaching Autobiography was that two readings assigned to broaden the prospective teachers' thinking were not shown to significantly affect most prospective teachers' thinking, nor were they reported to be useful by the prospective teachers. This finding supported streamlining the assignment, thus making time for other activities.

Example 2: Studying Broader Approaches to Mathematics Teacher Education

A study that focused on the effects of using a video case curriculum in the course provided the context for studying broader educational issues—in this case, the learning outcomes from using video-based materials in preservice teacher education (Stockero, 2008a, 2008b). After incorporating the materials in the course, the instructor/researchers sensed that class discussions were different and that the nature of the learning had changed, but there was no concrete evidence to describe the differences. A systematic analysis of discussion and written data from the course helped to better understand and articulate these changes.

This study documented a number of positive learning outcomes related to using the video case curriculum in the course, including richer reflections on practice, increased use of evidence to support analyses of teaching and learning, a more tentative stance toward analyzing practice, and an increased focus on student thinking. One particularly surprising finding was the way that prospective teachers came to analyze and consider various interpretations of individual student thinking, a sharp contrast to the generalities about student thinking that had previously dominated course discussions. These results affirmed the decision to replace prior course activities with the video case curriculum and provided a basis for thinking about how to integrate and sequence this curriculum and other course activities most effectively to capitalize on the learning that the video case curriculum had been empirically shown to support.

Example 3: Research on the Long-term Effects of Student Learning

The instructor/researchers are currently studying the extent to which prospective teachers' experiences and learning in the methods course have long-term effects on professional practice. This work, built directly on the video case study described above, questions whether observed learning outcomes are durable over time and the extent to which this type of learning later supports beginning teachers. To study these effects, a group of program graduates who used the video case curriculum in their methods course was convened to engage in activities parallel to those in the course; a subset of these graduates was later observed teaching.

Perhaps the most striking and unexpected finding in this work so far has been the durability of the norms for analyzing teaching practice established in the methods course. Without prompting, the graduates interacted with each other as they had interacted in the course: using mathematical arguments rather than procedural explanations, backing up claims with evidence, and pushing one another's mathematical and pedagogical thinking. The potential of these practices to support school students' mathematical learning (McClain & Cobb, 2001), as well as teacher learning (Grant, Lo, & Flowers, 2007), prompted the authors to begin to think about the importance of these norms in the methods course and what they can do to support prospective teachers in developing similar norms in their own classrooms.

What If Research Is Not Part of Your Job Description?

Although the authors engage in this work as researchers, they contend that systematic analysis of course design and instruction is a critical component of being a professional teacher educator. While reflecting on practice in informal ways has identified issues needing attention and ideas that could be tried out in the methods course, careful attention to data revealed nuances in prospective teacher learning and helped to identify specific course activities and instructor moves to more effectively meet course goals. Thus, other teacher educators are encouraged to consider incorporating systematic research into their work by using the tools of research found useful in supporting the learning of prospective teachers. By articulating problems in the practice of educating teachers, and then using such artifacts of practice as prospective teacher work and recordings of course discussions to study specific aspects of those problems, teacher educators can make more informed decisions about their own practice.

Conclusion

Teacher educators, as well as teachers in general, need to consider how research can be systematically incorporated into practice. Reading research results, thoughtfully considering how those results can be applied in practice, designing and testing new ideas and practices, and systematically studying the results enabled the authors to improve their practice and that of their students. This article highlighted the integration of research and practice by discussing the use of others' research to inform practice, ways in which prospective teachers are supported in developing habits reminiscent of using and conducting research, and how research can be used to answer questions about teacher educators' courses.

References

Ball, D. L., & Cohen, D. K. (1999). Developing practices, developing practitioners: Toward a practice-based theory of professional education. In L. Darling Hammond & G. Sykes (Eds.), *Teaching as the learning profession: Handbook of policy and practice* (pp. 3–32). San Francisco: Jossey-Bass.

Ball, D. L., Thames, M. H., & Phelps, G. (2008). Content knowledge for teaching: What makes it special? *Journal of Teacher Education, 59*, 389–407.

Cohen, E. G. (1998). Making cooperative learning equitable. *Educational Leadership, 56*, 18–21.

Cohen, E. G., Lotan, R. A., Scarloss, B. A., & Arellano, A. R. (1999). Complex instruction: Equity in cooperative learning classrooms. *Theory into Practice, 38*, 80–86.

Darling-Hammond, L., & Hammerness, K. (with Grossman, P., Rust, F., & Shulman, L.). (2005). The design of teacher education programs. In L. Darling-Hammond & J. Bransford (Eds.), *Preparing teachers for a changing world* (pp. 390–441). San Francisco: Jossey-Bass.

Grant, T. J., Lo, J.-J., & Flowers, J. (2007). Shaping prospective teachers' justifications for computation: Challenges and opportunities. *Teaching Children Mathematics, 14*, 112–116.

Lannin, J., Barker, D., & Townsend, B. (2006). Why, why should I justify? *Mathematics Teaching in the Middle School, 11*, 437–443.

McClain, K., & Cobb, P. (2001). An analysis of development of sociomathematical norms in one first-grade classroom. *Journal for Research in Mathematics Education, 32*, 236–266.

Silver, E. A., & Stein, M. K. (1996). The QUASAR Project: The "Revolution of the Possible" in mathematics instructional reform in urban middle schools. *Urban Education, 30*, 476–521.

Simon, M. A., & Tzur, R. (2004). Explicating the role of mathematical tasks in conceptual learning: An elaboration of the hypothetical learning trajectory. *Mathematical Thinking and Learning, 6*, 91–104.

Smith, M. S., Hughes, E. K., Engle, R. A., & Stein, M. K. (2009). Orchestrating discussions. *Mathematics Teaching in the Middle School, 14*, 548–556.

Smith, M. S., Stein, M. K., Arbaugh, F., Brown, C. A., & Mossgrove, J. (2004). Characterizing the cognitive demands of mathematical tasks: A sorting activity. In G. W. Bright & R. N. Rubenstein (Eds.), *Professional development guidebook for perspectives on the teaching of mathematics* (pp. 45–72). Reston, VA: National Council of Teachers of Mathematics.

Stein, M. K., & Smith, M. S. (1998). Mathematical tasks as a framework for reflection: From research to practice. *Mathematics Teaching in the Middle School, 3*, 268–275.

Stein, M. K., Smith, M. S., Henningsen, M. A., & Silver, E. A. (2000). *Implementing standards-based mathematics instruction: A casebook for professional development*. New York: Teachers College Press.

Stockero, S. L. (2008a). Differences in preservice mathematics teachers' reflective abilities attributable to use of a video case curriculum. *Journal of Technology and Teacher Education, 16*, 483–509.

Stockero, S. L. (2008b). Using a video-based curriculum to develop a reflective stance in prospective mathematics teachers. *Journal of Mathematics Teacher Education, 11*, 373–394.

Van Zoest, L. R., & Stockero, S. L. (2008a). Synergistic scaffolds as a means to support preservice teacher learning. *Teaching and Teacher Education, 24*, 2038–2048.

Van Zoest, L. R., & Stockero, S. L. (2008b). Using a video-case curriculum to develop preservice teachers' knowledge and skills. In M. S. Smith & S. Friel (Eds.), *AMTE Monograph 4: Cases in mathematics teacher education: Tools for developing knowledge needed for teaching* (pp. 117–132).

Van Zoest, L. R., & Stockero, S. L. (2008c). Concentric task sequences: A model for advancing instruction based on student thinking. In F. Arbaugh & P. M. Taylor (Eds.), *AMTE Monograph 5: Inquiry into mathematics teacher education* (pp. 47–58).

Van Zoest, L. R., & Stockero, S. L. (2009). Deliberate preparation of prospective mathematics teachers for early field experiences. In D. S. Mewborn & H. S. Lee (Eds.), *AMTE Monograph 6: Scholarly practices and inquiry into the mathematics preparation of teachers* (pp. 153–169).

Wood, T. (1998). Alternative patterns of communication in mathematics classes: Funneling or focusing? In H. Steinbring, M. G. Bartolini Bussi, & A. Sierpinska (Eds.), *Language and communication in the mathematics classroom* (pp. 167–178). Reston, VA: National Council of Teachers of Mathematics.

Laura R. Van Zoest is a professor of mathematics education at Western Michigan University. She studies the process of becoming an effective mathematics teacher and ways university coursework can accelerate that process. She can be reached at laura.vanzoest@wmich.edu

Shari L. Stockero is an assistant professor of mathematics education at Michigan Technological University. She is interested in understanding how to best support both prospective and practicing teachers' learning, particularly using practice-based materials. She can be reached at stockero@mtu.edu.

Alden J. Edson is a graduate student at Western Michigan University and a doctoral fellow with the Center for the Study of Mathematics Curriculum. He is currently interested in high school mathematics curriculum and interactive digital resources. His email is alden.j.edson@wmich.edu

van den Kieboom, L. A., and Magiera, M. T.
AMTE Monograph 7
Mathematics Teaching: Putting Research into Practice at All Levels
© 2010, pp. 173–189

12

Developing Preservice Teachers' Mathematical and Pedagogical Knowledge Using an Integrated Approach

Leigh A. van den Kieboom
Marquette University

Marta T. Magiera
Marquette University

This paper describes how an integrated mathematics content and early field-experience course provides opportunities for preservice elementary teachers to develop understanding of mathematics and mathematics teaching. Engaging preservice teachers in solving and discussing mathematical tasks and providing opportunities to implement these tasks with elementary students creates an authentic context for the future teachers to reflect on their own understanding of mathematics, mathematics teaching, and students' mathematical thinking. Essential elements of the cycle of events in the integrated model of instruction are discussed: preservice students' acquisition of mathematical concepts in the context of selected tasks in the content course; subsequent posing of mathematical tasks in early field experiences; reflection on work with students; and response to instructors' feedback.

The 2008 National Council on Teacher Quality (Greenberg & Walsh) report included five standards intended to guide reform efforts for the preparation of elementary mathematics teachers. The overarching theme of the standards was strengthening preservice elementary teachers' subject matter knowledge. The main recommendation for design of coursework for prospective elementary teachers focused on their "unique needs," emphasizing the ways in which they need to know and understand elementary mathematics. This unique kind of understanding is frequently described by the mathematics education community as specialized mathematical knowledge for teaching, and comprises (broadly defined) mathematical and pedagogical content knowledge (Ball & Bass, 2003; Hill, Ball, & Schilling, 2008).

Preservice teachers typically arrive at a university with a procedural understanding of elementary mathematics and a strongly held belief that procedural understanding is the core of mathematics learning. Such understanding and belief, having developed through 12 years of procedure-oriented mathematics instruction, interferes with preservice teachers' abilities to acquire subject matter knowledge in a meaningful way (Ball, 1990). Strong subject matter knowledge is an important requisite for establishing pedagogical content knowledge (Capraro, Capraro, Parker, Klum, & Raulerson, 2005). Therefore it stands to reason that preservice teachers typically fail to acquire an understanding of the pedagogical components needed to teach mathematics. For example, Crespo (2003) and Crespo and Sinclair (2008) document that preservice teachers have limited abilities to select and pose good mathematical tasks that engage students in thinking about mathematics. Nicol (1999), and Moyer and Milewicz (2002) draw attention to the fact that preservice teachers' questioning skills are inadequate to probe students' understanding and move them beyond providing an answer to a problem.

Needed pedagogical skills develop slowly over time in mathematics coursework when preservice teachers explicitly engage in analysis, discussion, and reflection on students' mathematical thinking given the support and guidance of mathematics educators. Thus, effective teacher preparation programs need to provide future teachers with compelling opportunities to acquire and strengthen both components of mathematical knowledge for teaching: mathematics content and pedagogical content knowledge.

Completing appropriate college mathematics and methods courses does not necessarily guarantee that preservice teachers will use what they learn to inform their work with students in field experiences, student teaching, or beginning practice. Borko et al. (1992) provide evidence of how a student teacher who had finished a significant number of college mathematics courses struggled to explain to a sixth grade class why and how the standard algorithm for dividing fractions works. Ebby (2000) argued that unless preservice teachers learn how to make direct connections between the mathematics they learn in their content courses with what they learn about teaching mathematics in methods courses and field experiences, teacher preparation programs will remain a weak intervention. As such, programs might fail to change the beliefs and the effect of the experiences that preservice teachers have as they begin their studies.

This paper explores how integrating mathematics and pedagogy in the context of *early* field experiences (prior to student teaching) induces preservice teachers to develop knowledge of mathematics and teaching mathematics that supports student learning. The authors demonstrate how interactions between teaching and learning and between knowledge and practice provide preservice teachers with authentic opportunities to analyze students' thinking and reflect on their own teaching actions. Ebby (2000) emphasized that such opportunities are essential to help preservice teachers internalize different aspects of mathematical knowledge for teaching.

Bridging Mathematical and Pedagogical Content Knowledge in the Early Field Experience: An Integrated Model of Instruction

The sequence of integrated mathematics courses discussed here was designed using recommendations from the Conference Board of the Mathematical Sciences (CBMS, 2001) as a framework. The purpose was to provide preservice teachers with opportunities to make direct connections between mathematics and pedagogy by linking learning in mathematics content courses to direct work with students in early field experiences.

The integrated model provided preservice teachers with authentic opportunities to reflect on their personal knowledge and practice.

Taught jointly by faculty from the Department of Mathematics, Statistics, and Computer Science and the College of Education, the integrated sequence consisted of two courses: (1) Number Systems and Operations, and (2) Algebra and Geometry for Teachers. An early field experience was integrated into each course. Preservice teachers were provided opportunities to strengthen mathematical knowledge for teaching by implementing selected mathematical tasks with elementary students in early field experiences.

The mathematics and pedagogy embedded in each task were first discussed in content courses. Then, the preservice teachers, under the supervision of the course instructor, implemented the selected tasks with students in the early field experiences. Working directly with students provided preservice teachers opportunities to analyze elementary students' mathematical thinking, reflect on their own understanding of these same concepts, and reflect on their own teaching actions. In addition, the integrated course sequence created opportunities for course instructors to continuously assess preservice teachers' mathematical and pedagogical knowledge, allowing for individualized support and intervention. To illustrate how the integrated model supports preservice teachers' learning of mathematics and pedagogy the authors use examples from the Number Systems and Operations course, specifically the fractions unit.

The Study of Fractions

For preservice teachers and elementary students, understanding fractions is one of the more difficult topics in the elementary mathematics curriculum (e.g., Lamon, 2007; Ma, 1999; Newton, 2008). Lamon (2007) argued that difficulties with understanding and teaching the concept of fractions relate to the complexity of fraction representations. Kieren (1976) originally emphasized the complex nature of fractions by identifying four different subconstructs for interpreting the meaning of fractions: ratio, measure, operator, and quotient. Each interpretation builds on the part-to-whole relationship (Behr, Harel, Post, & Lesh, 1992). In fractions literature, the part-to-whole subconstruct is defined as a comparison of one or more equal parts of a unit to the total number of equal parts into which a unit is divided. The ratio subconstruct expresses a part-to-part comparison of two quantities where the number of units in the first quantity relates to the number of units in the second quantity. The measure subconstruct represents the notion of density on the number line, emphasizing the role of unit fractions and fostering knowledge of fractions as additive quantities. The operator subconstruct supports acquisition of multiplicative reasoning. The quotient subconstruct employs two different interpretations of fraction division: partitive (how many in each group) and quotative (how many groups). The unit on fractions included in the Number Systems and Operations course utilized the different subconstructs to assist preservice teachers in developing an understanding of fractions.

Selecting Mathematical Tasks

To provide preservice teachers opportunities to develop a complex and deep understanding of fractions and to examine their own mathematical and pedagogical knowledge, the authors selected tasks to be used by the preservice teachers and their field students. The Mathematical Tasks Framework (Stein, Smith, Henningsen, & Silver, 2000) and descriptions of worthwhile mathematical tasks (National Council of Teachers of Mathematics, NCTM, 1991) guided task selection. These frameworks provided a filter for selecting tasks with potential to move preservice teachers and their elementary students from a procedural understanding to a conceptual understanding of fractions. Selected tasks had the potential to elicit problem solving, reasoning, communication, and making connections in order to help preservice teachers build an understanding of the meaning of fractions and operations with fractions using the four subconstruct models; learn the pedagogical content knowledge related to student misconceptions about fractions; explore different materials available for teaching and learning about fractions; and practice a variety of teaching strategies. In addition, selected tasks transferred to the field experience as viable problems for elementary students to solve. Example tasks are shown in Figure 1.

Subconstruct	Task Number	Mathematical Task
Part-to-Whole	1	[1]Kayla says that the shaded part of the picture can't represent ¼ because there are 3 shaded circles and 3 is more than 1, but ¼ is supposed to be less than 1. What can you tell Kayla about fractions that might help her?
Ratio	2	Andy and his sister Amy are making lemonade for their lemonade stand. Which of the following two mixtures will make the lemoniest lemonade? Mixing three tablespoons of lemon juice with four cups of water or mixing four tablespoons of lemon juice with five cups of water? Use as many ways as you can think of to solve this problem. Each time, clearly explain your thinking.

[1] Tasks adapted from Beckmann, S. (2008). *Mathematics for elementary teachers with activities manual* (2nd ed.). Boston, MA: Pearson.

Subconstruct	Task Number	Mathematical Task
Operator	3	Demarco used 3/4 cup of cheese in the pan of lasagna he made. His younger brother Anthony ate 5/16 of the pan of lasagna. What fraction of a cup of cheese did Anthony consume when he ate the lasagna? Use area drawings to show how you solved the problem? Explain how your drawings helped you to solve it.
Quotient	4	Mary has 3-1/4 yards of fabric to make dresses for her dolls. Each dress requires 2/3 of a yard of fabric. How many dresses can she make? Will she have any fabric left? How much? Use a drawing to solve the problem.
Measure	5	Locate 15/24 on the number line 0 $\frac{1}{4}$

Figure 1. Examples of mathematical tasks used in the fraction unit.

Data presented in the next section comes from transcriptions of videotaped preservice teachers' interactions in the content class and audiotapes of preservice teachers' interactions with field students, as well as reflective journals. These data were part of a larger project that followed 27 preservice teachers from their content class to their early field experience.

Solving Mathematical Tasks in the Content Course

To stimulate their thinking about fractions as a relationship between part-to-whole and part-to-part, the preservice teachers worked in the content class in small groups on tasks similar to Task 1 (see Figure 1). The transcription below illustrates a discussion as the preservice teachers shared their thinking, anticipating various ways elementary students might reason while solving these types of tasks.

Instructor: How would you explain that this [referring to the picture in Task1] represents one fourth? Karen?

Karen: I think there would be two ways to do it. One way was if you put it into four. If you put a box around all the four different groups of three circles and then showed it as each cluster is one part.

Instructor: Do you want a big box around all of those? [referring to the picture in Task1]

Karen: Well, around each. You can make one bar and then separate each group of three [instructor draws a vertical line between each collection of three circles to separate each group]. Then you could see that as one fourth. Then I thought another way you could do it, is you counted, I don't know if this makes it more difficult, but if you counted all of the circles and you made it to three over twelve and then you could reduce it to one fourth. But I don't know if that would be too difficult.

Instructor: Okay, those are both good ideas. Somebody want to add something to that?

Carrie: You could explain to the kids that, one little circle is not the whole, in this case, the whole is, all the circles together.

Instructor: So, what's really important here is that you define the whole. So the whole is twelve circles, right? Once I know that, then I can say that this is three out of twelve. If they say that's one third [pointing to the three shaded circles], what are they thinking about [this situation] if they think that [the picture] represents one third rather than one fourth? One group of three is shaded and, how many are not shaded? Three groups of three. So, they're really thinking, this part to this part [pointing at one group of shaded circles and three groups of unshaded circles], and actually, that's a ratio. So, fractions are part to whole, you have to know what the whole is. And the whole is twelve circles.

Gina: I was thinking of it in terms of groups of shaded and unshaded circles. It's three. Some kids think that if it's three, that's thirds, so, but I don't know? But I am … Is this reciprocal thinking?

Instructor: That's interesting, I never thought of it that way. So the reciprocal of three is one third, that's true. But, I don't think that's what kids are thinking when you ask them what fraction of the circles is shaded, and they say one third. One group of three is shaded and how many are not shaded? Three groups of three. So, they're really thinking, this part to this part [pointing at shaded and unshaded groups of circles], and actually, that's a ratio. When you do part to part, all right? So, fractions as part to whole, you have to know what the whole is. And the whole is twelve circles. So it is very deceiving—you can see three out of twelve or one to three.

The mathematical task provided a context for preservice teachers to consider different interpretations of fractions. They engaged in a discussion about the part-to-whole subconstruct. Karen's contributions indicated two different views of the whole: a collection of four groups of three circles, and a collection of 12 circles. Karen's first approach identified one-fourth directly, as one group of three shaded circles out of four groups of three circles. Karen's second approach focused class discussion on interpreting three-twelfths as one-fourth, indicating a different view of the whole, a collection of 12 individual circles.

The preservice teachers also considered different kinds of pedagogical content knowledge needed to implement this task with elementary students. Discussion created an opportunity to examine and reflect on possible students' interpretations and misconceptions about the meaning of fractions. For example, Carrie emphasized that a teacher needed to discuss the meaning of the whole while working with students. Gina pointed out that students might focus on the relationship between groups of shaded and unshaded circles, providing an opportunity for another discussion of students' misconceptions. In addition, during class discussions preservice teachers considered various materials to support students' thinking about fractions and various questions they might pose during the early field experience.

Posing Mathematical Tasks for Students in the Early Field Experience

Each week during the early field experience preservice teachers worked with a classroom teacher, assisting the teacher in conducting a 60-minute mathematics lesson. Then each preservice teacher worked directly with two students from the classroom, conducting a 30-minute activity session. The activity sessions provided the preservice teachers with opportunities to pose selected mathematical tasks for their students. Each session was audiotaped and observed by an instructor. After completing each session, preservice teachers reviewed the audiotape and reflected on their teaching actions, as illustrated by the transcript excerpt, which documents Karen's interactions with a student while she implemented Task 1.

Student: [reading the problem] Kayla says that the shaded part of the picture can't represent one fourth because there are three shaded circles and three is more than one but one fourth is supposed to be less than one. What can you tell Kayla about fractions that might help her?

Karen: So, how do you think she got that? She said the shaded part of the picture can't be one fourth because there are three shaded circles and three is more than one.

Student: So it can't be one third.

Karen: One third?

Student: All three of them are colored.

Karen: If you just looked at that picture, what does it show you?

Student: Okay, I know that it is one fourth.

Karen: Four groups? Or four, just four circles?

Student: Well it's four circles, no, four groups.

Karen: Four groups. Okay. And then, so what is that? Is that our numerator or denominator?

Student: Denominator.

Karen: Which one? What do we say all the pieces make?

Student: Denominator.

Karen: Denominator, right. So, if we have four groups, that makes our whole. What's our numerator?

Student: One.

Karen: Why is it one?

Student: Because, the one that she shaded in, she shaded in one group out of four.

Karen: Okay, so that's one-fourth. How did she get three? It says that the picture can't be one fourth because there's three shaded circles.

Student: And one group is three circles and she shaded the three circles out of one group so that's how she got three.

Karen: Out of one group? So that's how she got that?

Student: Hm-hm.

Karen: But, we know that's not right? Because we see that there's four groups, right?

Student: Yeah. It's four groups, but she took one group and shaded three things out of one.

Karen: Right. So, our fraction right there is one-fourth, right?

Student: Yes.

The transcript excerpt shows that Karen guided the student toward the answer by posing leading questions rather than probing the student's understanding of a fraction, a ratio, and the difference between the two. She failed to provide a full explanation for the difference between part-to-whole and part-to-part constructs embedded in the task. She used the terms *whole*, *numerator*, and *denominator* explicitly throughout her work with the student, without connecting these terms to the picture in the problem or to the symbolic notation for the fraction $\frac{1}{4}$. She failed to make connections with the picture and build the meaning of one-fourth and three-twelfths based on what the student said and thought about the fraction. It is not clear what the student knew and understood about the meaning of fractions.

Karen's reflection on her interactions with the student reveals her lack of awareness of her limited mathematical and pedagogical knowledge. Instructors provided feedback and intervention through individual conferences and follow-up discussions in the content course to address Karen's deficiencies and help her strengthen and link her mathematical and pedagogical knowledge.

> I used Kayla's problem (Task 1) to help me explain to the students that a fraction is a part-to-whole relationship not a ratio or part-part relationship. This problem called for students to take a look at the group of shaded circles in comparison to the other circles. For this, they had to understand that this one group of three was one shaded group out of four groups of three circles each, and therefore, [one-fourth]. This problem, I feel, clarified the idea of part to whole relationships. It helped me find new ways of explaining these concepts to the students. It gave me new ways of looking at normal fraction problems and gave me the confidence I needed to be able to teach these to my students. Basically it provided a framework of thought that helped me look at math in the perspective of a teacher trying to get a point across, rather than a student finding answers. Now I have both perspectives (both teacher and student) to help me find ways to better tutor my students.

Feedback focused on mathematics and pedagogy and emphasized missed opportunities to probe elementary students' thinking about the whole, to clarify the part-to-whole meaning of fractions, and to build on student's ideas that could possibly lead to part-to-part interpretation. Instructor feedback created the opportunity for Karen to re-examine her mathematical and pedagogical content knowledge to focus on students' learning and classroom instruction prior to returning to the field. The integrated model created for Karen a sustained cycle of learning, teaching, and reflecting on her own practice.

Conclusion and Recommendations

Each element in the cycle of events described in this paper—discussing mathematical concepts in the context of selected tasks in the content course, selecting and posing mathematical tasks in the early field experience, reflecting on work with students, and responding to instructors' feedback—engages preservice teachers in a dialogue about the teaching and learning of mathematics that contributes to the development of their mathematical knowledge for teaching. As illustrated in Figure 2, the integrated model of instruction provides a way for preservice teachers to examine the connections between mathematical and pedagogical knowledge.

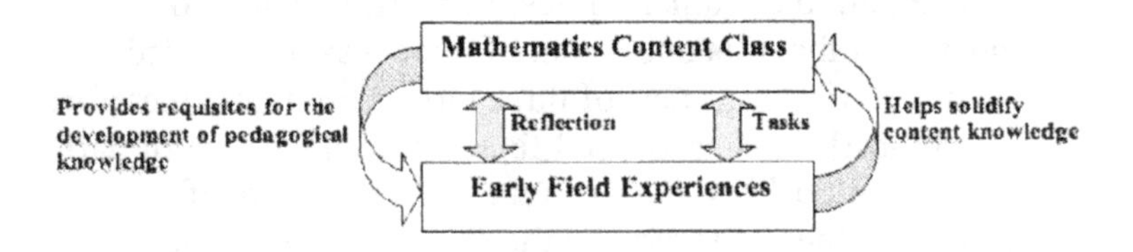

Figure 2. Developing mathematics knowledge for teaching in the context of integrated instruction.

The integrated instruction model gives preservice teachers authentic opportunities to connect their learning of mathematics with their learning about how to teach mathematics in practice. The mathematical tasks serve as a bridge linking preservice teachers' learning of mathematics and pedagogy. There are many mathematics topics that preservice teachers do not experience in this way. More work is needed to identify and to develop mathematical tasks to help preservice teachers examine mathematical concepts, elicit their thinking about how to teach these concepts, and heighten awareness of students' mathematical thinking and learning. These tasks must support the interrelated goals of strengthening preservice teachers' mathematical and pedagogical knowledge and at the same time be viable problems for elementary students to solve in the early field experience.

References

Ball, D. L. (1990). The mathematical understanding that prospective teachers bring to teacher education. *Elementary School Journal, 90*, 449–466.

Ball, D. L., & Bass, H. (2003). Toward a practice-based theory of mathematical knowledge for teaching. *Proceedings of the 2002 Annual Meeting of the Canadian Mathematics Education Study Group* (pp. 3–14). Edmonton, AB: CMESG/GCEDM.

Behr, M. J., Harel, G., Post, T., & Lesh, R. (1992). Rational number, ratio, and proportion. In D. Grouws (Ed.), *Handbook of research on mathematics teaching and learning* (pp. 296–333). Reston, VA: National Council of Teachers of Mathematics.

Borko, H., Eisenhart, M., Brown, C. A., Underhill, R. G., Jones, D., & Agard, P. C. (1992). Learning to teach hard mathematics: Do novice teachers and their instructors give up too easily? *Journal for Research in Mathematics Education, 23*, 194–222.

Capraro, R. M., Capraro, M. M., Parker, D., Kulm, G., & Raulerson, T. (2005). The mathematics content knowledge role in developing preservice teacher's pedagogical content knowledge. *Journal of Research in Childhood Education, 20*, 102–118.

Conference Board of the Mathematical Sciences. (2001). *The mathematical education of teachers.* Providence, R.I. and Washington, DC: American Mathematical Society and Mathematical Association of America.

Crespo, S. (2003). Learning to pose mathematical problems: Exploring changes in preservice teachers' problems. *Educational Studies in Mathematics, 52*, 243–270.

Crespo, S., & Sinclair, N. (2008). What makes a problem mathematically interesting? Inviting prospective teachers to pose better problems. *Journal of Mathematics Teacher Education, 11*, 395–415.

Ebby, C. B. (2000). Learning to teach mathematics differently: The interaction between coursework and fieldwork for preservice teachers. *Journal of Mathematics Teacher Education, 3*, 69–97.

Greenberg, J., & Walsh, K. (2008). No common denominator: The preparation of elementary teachers in mathematics by America's education schools. Retrieved from National Council on Teacher Quality website: www.nctq.org

Hill, H. C., Ball, D. L., & Schilling, S. G. (2008). Unpacking pedagogical content knowledge: Conceptualizing and measuring teachers' topic-specific knowledge of students. *Journal for Research in Mathematics Education, 39*, 372–400.

Kieren, T. (1976). On the mathematical, cognitive and instructional foundations of rational numbers. In R. Lesh (Ed.), *Number and measurement: Papers from a research workshop* (pp. 101–144). Columbus, OH: ERIC/SMEAC

Lamon, S. J. (2007). Rational numbers and proportional reasoning: Toward a theoretical framework for research. In F. K. Lester (Ed.), *Second handbook of research on teaching and learning mathematics* (pp. 157–223). Reston, VA: National Council of Teachers of Mathematics.

Ma, L. (1999). *Knowing and teaching elementary mathematics: Teachers' understanding of fundamental mathematics in China and the United States*. Mahwah, NJ: Erlbaum.

Moyer, P. S., & Milewicz, E. (2002). Learning to question: Categories of questioning used by preservice teachers during diagnostic mathematics interviews. *Journal of Mathematics Teacher Education, 5*, 293–315.

National Council of Teachers of Mathematics. (1991). *Professional standards for teaching mathematics*. Reston, VA: Author.

Nicol, C. (1999). Learning to teach mathematics: Questioning, listening, and responding. *Educational Studies in Mathematics*, *37*, 45–66.

Newton, K. J. (2008). An extensive analysis of preservice elementary teachers' knowledge of fractions. *American Educational Research Journal*, *45*, 1080–1110.

Stein, M. K., Smith, M. S., Henningsen, M. A., & Silver, E. A. (2000). *Implementing standards based mathematics instruction: A casebook of professional development*. New York: Teachers College Press.

Leigh A. van den Kieboom is an assistant professor in the College of Education at Marquette University. Her research focuses on the content and pedagogical knowledge preservice teachers need to develop prior to practice. She can be reached at leigh.vandenkieboom@marquette.edu

Marta T. Magiera is an assistant professor of mathematics education in the Department of Mathematics, Statistics and Computer Science at Marquette University. Her research interests include mathematical thinking, metacognition, problem solving, and mathematics teacher development. She can be reached at marta.magiera@marquette.edu

Benken, B. M., and Gomez-Zwiep, S.
AMTE Monograph 7
Mathematics Teaching: Putting Research into Practice at All Levels
© 2010, pp. 191–205

13

A Case Study of Alternative Pathways to Secondary Teacher Certification in Mathematics: How Should the Pathways Be Structured?

Babette M. Benken
California State University, Long Beach

Susan Gomez-Zwiep
California State University, Long Beach

A decade of teacher shortages in California has prompted the State Board of Education to offer alternative options for individuals seeking a secondary teaching credential. Although alternative programs have emerged rapidly, little is known as their nature, extent, and effectiveness. This study examines one such program that provided secondary alternative setting teachers with an expedited pathway towards additional certification in mathematics through content and methodology coursework aimed at preparing participants to pass state exams. During the program, participants expanded content understandings, as well as beliefs about and attitudes toward mathematics and its teaching and learning. While most participants were confident in their ability to pass the exams, they did not yet feel adequately prepared, and struggled to complete preparation independently. Suggestions for how to structure and support alternative certification programs are provided.

The ongoing teacher shortage over the past decade in California has prompted the State Board of Education to offer alternative options for individuals seeking a secondary teaching credential (Chin & Young, 2007). One option, the California Subject Examinations for Teachers (referred to as *Exam Option*) (2009), allows individuals to satisfy the subject matter portion of their credentials via a set of exams, rather than completing a formal major in the discipline. In mathematics, there are two pathways: (a) passing Subtest I (Algebra and Number Theory) and Subtest II (Geometry, Probability, and Statistics) yields a secondary credential in "Foundational Mathematics" (allowing teaching through second year algebra), or (b) passing Subtests I, II, and III (where subtest III includes Calculus and History of Math) yields a full secondary credential (allowing teaching of any mathematics course for grades 7–12).

The United States Department of Education (2002, 2004) has repeatedly claimed that alternative route (as opposed to traditional) programs are an effective means of streamlining the process of certification to move teachers quickly into the classroom, and federal legislation (U. S. Congress, 2001) has encouraged these approaches. Although alternative programs have emerged at a rapid rate, with over 200,000 persons receiving certification through such programs by 2004 (Feistritzer, Haar, Hobar, & Scullion, 2005), little is known as to the nature, extent, and effectiveness of these programs (Rosenberg & Sindelar, 2005). As Zientek (2007) noted, "[Alternative certification programs] have established their place in education. Therefore, the focus should shift from whether they should exist to ensuring effective teacher preparation programs are created and high quality teachers are produced" (p. 960). While this approach immediately provided districts with teachers, a concern has been that many certified in this manner were neither prepared to teach the content in ways that facilitated conceptual understanding (Hill, Rowan & Ball, 2005), nor felt ready for instructional settings (Zientek, 2007).

A Case Study

A team of faculty from a large, urban state university and a local county office of education engaged in a grant-supported program that supported a cohort of teachers of grades 6-12 working in juvenile hall facilities to garner the subject matter and teaching knowledge needed to pass the Exam Option. The overarching goal of this program was to provide teachers in these facilities with an expedited pathway to gain additional certification in mathematics and/or science. Although the 32 teachers in this program already held elementary or secondary credentials in other disciplines, all were interested in obtaining a secondary credential in mathematics or science, as many were teaching classes beyond their credentials. Most mathematics cohort participants had taken few recent mathematics courses, supporting LaTurner's (2002) observation that as of 2002, almost 50% of all mathematics and science teachers in grades 6-12 across the nation were teaching without the necessary qualifications.

This program included four courses (three for mathematics or science content, one for blended mathematics/science methods) spread over 1.5 years (one course each semester, including summer). All courses were taught by tenure-line faculty and were held at the participants' county education office in a classroom setting. In an attempt to explore what is possible with this alternative approach to certification, the following questions guided a research study of the 24 participants who completed the mathematics pathway: In what ways did participants appear to expand their beliefs and content understandings? How should alternative pathways be structured to facilitate teacher learning in mathematics? Data sources included: (a) participant information/belief surveys (pre and post tests for each course), (b) semi-structured interviews with course instructors and a subset of participants (end of program), (c) content knowledge exams (pre and posttests for each course), (d) course observation notes, (e) participant scores (grades and exam scores), (f) program artifacts (e.g., syllabi), and (g) researcher journals. The surveys included open-ended as well as Likert-type questions ranging from 1 = strongly disagree to 6 = strongly agree. Sample open-ended questions included: *What are your expectations for this course? In general, do you feel you are good at math? As a result of this course, do you feel more confident in your ability to pass the state exams?* Sample scaled statements included: *Mathematics involves mostly facts and procedures that have to be learned,* and *in mathematics, answers are either right or wrong*. The survey questions helped to quantify trends and growth in participants' knowledge of and beliefs about mathematics and its teaching and learning.

In interviews participants were asked about their learning experiences, their preparedness for and success in the Exam Option tests and future teaching practice, and the alternative nature of the program. Instructors were similarly asked about their experiences, their perceptions of participants' preparedness for the Exam Option tests, and suggestions they had for the program and its structure based on their experience.

Data analysis was two-tiered, involving individual case studies and an over-arching study of the entire cohort. Based on their frequency, agreement, and level of importance, themes were developed from detailed analysis through a systematic process of coding (Stake, 1995). Some themes were developed through the analysis process. Others were anticipated as analysis began: these included knowledge and beliefs of participants related to their experiences in the program, their mathematical content knowledge, and confidence in their abilities to teach that content. Quantitative analyses produced descriptive statistics of program completion rates, course grades, and results from Likert-type survey questions. Reliability was addressed by using two researchers to code statements, and validity was enhanced through triangulation of data with researcher notes.

Framework for Content Courses

Teachers' thinking about mathematics teaching and learning is challenged by the expectations and ideals endorsed by reform (National Council of Teachers of Mathematics, 2000). Such recommendations call for an approach to teaching that allows students to communicate, solve problems, and engage in conceptually based content. Teachers are asked to teach in ways that promote an integrated, connected view of content, rather than a procedural, rule-based view. In this alternative route program, teachers' understanding of content was central. Content embedded in the coursework paralleled that in the state examinations; however, due to time constraints, the courses did not attempt to address test preparation outside of including the content contained in the exams and exploring that content through some similar questions (e.g., multiple choice, open response). An outline of topics for each content course is provided in Table 1. This approach to instruction attempted to provide teachers with a conceptual understanding that would bridge to effective teaching.

Table 1
Content Course Topics

Course	Content
Course I (Fall)	Number systems/sets of numbers, number theory/set theory, algebra (e.g., linear, quadratic, exponential functions), proof (e.g., contradiction, induction), linear algebra, abstract algebra
Course II (Spring)	Geometry/geometric proof, right triangle trigonometry, transformations, constructions, probability (e.g., finite, conditional), statistics (e.g., interpretive, regression)
Course III (Summer)	Functions (e.g., trigonometric, polynomial), limits and continuity, derivatives, integrals, applications, sequences and series, history of mathematics

Knowing that teachers are unlikely to make adjustments in their thinking without intervention and deliberate support, professional development efforts must intentionally provide experiences to assist teachers in learning new ways of thinking about mathematics and its teaching (Farmer, Gerretson, & Lassak, 2003; National Mathematics Advisory Panel, 2008). Throughout the content courses, content was explored in ways that allowed teachers to develop conceptual understandings, particularly of foundational concepts (e.g., functions) that they themselves could be teaching in future practice. Opportunities were provided for teachers to explain their thinking about the content in both whole- and small-group settings. Problem solving and approaches to proof and justification were scaffolded by instructors, and connections within the disciplines were explicitly presented and explored.

When possible, content instructors also modeled effective practice and discussed how the teachers' new content understandings translated to effective secondary teaching. Ball, Lubienski, and Mewborn (2001) cite the importance of *knowing mathematics for teaching*, which encompasses all of the knowledge required to teach mathematics effectively. From a professional development standpoint, this perspective suggests that programs should provide opportunities to learn mathematics around specific content and teaching situations arising in practice. Additionally, the courses explicitly confronted teachers' beliefs about mathematics and addressed its teaching and learning simultaneously. Research suggests that teachers' knowledge of and beliefs about content are related in powerful ways (Wilson & Cooney, 2002); thus, courses need to address all aspects of the learners' cognition (e.g., beliefs, knowledge, affect) in order to facilitate the integration of new understandings.

The Methods Course

The methods course blended mathematics and science to emphasize pedagogical approaches common to both disciplines and addressed issues particular to alternative educational settings similar to those in which the participants taught. Most assessments were project-based, allowing participants to focus on either mathematics or an area of science. For example, one assignment required participants to analyze curriculum materials in light of student learning outcomes; in particular, they were asked to consider the relationship among the concepts, materials, and pedagogical strategies implicit in the lesson plan that contained them. As in the content courses, teachers' content knowledge and beliefs about content and teaching were explicitly addressed.

It is worth noting that the decision to offer one blended methods course was based on two pragmatic considerations. First, some teachers were attempting to complete both the mathematics and science programs and did not have time to take two methods courses. Second, not all teachers were required by the state to take a secondary methods course because they held a secondary credential in another discipline, and this resulted in very small numbers of students to take separate courses. While this format proved to be pragmatic and supported the expedited nature of this program, it was difficult finding an instructor who had the background for, and interest in, teaching such a blended course. Developers established three primary criteria essential for the instructor of this course: secondary teaching experience in mathematics or science, content expertise in both mathematics and science, and experience facilitating teacher development (preservice or inservice). An instructor was found who exceeded the criteria, having K–12 teaching experience in both disciplines and having led professional development for both secondary mathematics and science teachers.

Overview of Findings–Course Sequence

During the content and methods courses, teachers expanded their beliefs about, and attitudes toward, mathematics and its teaching and learning. Interestingly, the change was noticeably greatest in the first content course. The pre-survey showed that when beginning the program, attitudes toward the content courses were mixed; these feelings related directly to teachers' confidence in themselves relative to their content background and ability to learn that content. Those participants who had not taken many undergraduate level mathematics courses or who had not studied mathematics in many years were "nervous" or "uncertain" about taking the courses. Many exhibited symptoms of math anxiety (Hembree, 1990) and openly discussed their apprehensive feelings. However, those participants who perceived themselves to be "good at math" were "excited" about once again studying mathematics.

Following the series of content courses, all participants believed they had either learned new mathematics or reviewed essential mathematics for passing the Exam Option. While almost all (89%) felt more confident about passing the exams, many noted that they still had much to do to prepare. Thus, in their estimations they were *more* prepared, but not yet *adequately* prepared. Many participants commented on surveys/ interviews that the courses covered too much content, causing them to feel overwhelmed at times. On the course post-survey, some participants suggested that either more time was needed to explore all content adequately or fewer concepts should have been addressed. In spite of their concern over completing preparation for the Exam Option, most participants stated that they felt confident they could finish the preparation on their own. As one noted, "I have a better idea what I don't know, and can plan better how to remediate." Participants who had been nervous about being learners of mathematics were now ready to continue learning independently.

Overall, participants' knowledge of mathematical concepts and skills increased, as measured through pre and post exams, course artifacts, and instructors' observations. For example, during the first mathematics course (see Table 1), the average increase in score from pre to post test was 26%, with two participants' scores increasing by over 50%. Additionally, all students were able, by the end of the courses, to better communicate their thinking and understanding orally and in writing on quizzes and assignments. While it would be expected that participants would demonstrate an increase in knowledge following a content course, less anticipated was how quickly participants grew in solving problems, justifying responses, and articulating their thinking using appropriate terminology and in both small- and whole-group formats.

Not all participants were successful at passing the Exam Option exams. Of those attempting all three exams, only 21% (five participants) passed all of the exams, thereby providing them with a full secondary credential; 38% (nine participants) obtained foundational secondary certification.

As stated previously, the primary focus of the program's development was on the content courses, as participants needed the most help in building content knowledge to pass the exams. Since all participants were practicing teachers, the developers were not as concerned with their knowledge of general pedagogy or with the need for them to successfully complete a required methods course. Despite having approached the program with these assumptions, data revealed that participants did gain new pedagogical expertise within mathematics, and believed that the methods course was critically important to their future instruction of this new content. Additionally, participants noted on post surveys that the methods course helped them to recognize teaching strategies modeled in content courses. Finally, overt connections in the methods course to strategies for teaching in alternative settings helped participants to envision how new learning could be applicable in their current work environments.

Structure of the Program–Lessons Learned

What was learned relative to structuring alternative pathways to certification pertains to three main aspects of the program: (a) configuration of the professional development program, (b) support provided by the partners, and (c) expectations communicated to participants.

Configuration of the program. Both participants and course instructors believed that addressing the content contained on the certification exams in three content courses was not adequate given the background knowledge of participants. Most of the participants were not teaching mathematics, and many had not studied this content in over a decade. Additionally, most participants had taken very few applicable content courses; for example, over half of the mathematics participants had not taken courses beyond second year algebra or pre-calculus. As a result, course instructors had to revise syllabi numerous times in order to adjust the pacing of topics to meet the needs of learners.

Instructors expressed frustration over not having time to sufficiently address planned topics. For example, within the first content course the instructor did not have time to formally explore abstract algebra, which is covered on the state exam, as more time was needed with beginning algebra concepts (e.g., graphing). The authors recommend a minimum of four content courses to explore content for a subject matter competency exam required for certification; the additional time would allow for growth in both attitudes toward content and the ability to teach that content. If possible, it would also be helpful to create multiple smaller sections of a course in order to group participants based on their existing content understanding.

Holding classes at a county site became both a benefit and impediment to participant learning. The participants appreciated the close proximity to work, but they had difficulty meeting with instructors for office hours; classes usually began soon after participants finished the teaching day. Additionally, access to technology at times became difficult and unreliable, often forcing instructors to rework planned lessons. Although holding professional development events on-site can facilitate learning through convenient access and creation of a familiar professional learning community (Benken & Brown, 2008), the expedited nature of alternative pathway certification programs can sometimes also benefit from a more traditional environment that offers predictability in regards to access to pedagogical tools and instructors.

Support provided by the partners. The K–12 partner in an alternative certification program (in this context, the county office of education) must provide support in two areas. First, the partner should identify an on-site contact liaison. All instructors in this program frequently utilized the county liaison to help with participant communication, manage supplies, schedule courses/rooms, and complete daily office tasks. Second, systemic scaffolds are needed. Engaging in an expedited, content-focused program requires a large and intensive time commitment by participants. Some participants did not complete the program (or dropped from attempting both mathematics and science to completing one or the other) due to other outside commitments. This program needed support from the partner districts in terms of classroom release time, professional development time, or release from committee work to provide additional time for studies. For example, a periodic half-day release from teaching and committee duties for all participants would have allowed them to establish on-site learning communities as well as time to meet with instructors on campus.

It is equally important that the university as the higher education partner provide support, particularly relative to staffing courses. In this program, the university faculty needed support to teach the alternatively structured courses. Compensation for off-site locations, help with transporting supplies, and release time to teach alternatively offered courses are all useful, and often needed, supports. Although the entire sequence of courses in this program was staffed with full-time tenured or tenure-track faculty, such a scenario is not always realistic. Part-time faculty or individuals from outside the program may need additional time or support to understand its inherent complexities and issues.

Finally, the selection of instructors is critical to the success of such a program. Instructors must have the appropriate knowledge background but also be prepared to teach in alternative settings. Instructors need to account for the high level of math anxiety of participants as well as the logistical challenges of offering off-site courses. In an interview, the Course I instructor noted that most participants were not able to find time to commute to the university, and therefore on-site office hours were held before and after class.

Expectations communicated to participants. In expedited pathways to certification, the level of commitment required is immense. In order to facilitate successful completion by participants it is critical that they understand this enormous requirement. Both partners must communicate clear expectations about the load as part of recruitment, and participants must structure necessary personal time before beginning. Programs must only allow participants to embark on pathways for which there is reasonable expectation of completion. In this study, participation was allowed in both the mathematics and science cohorts; after the first content course, the number of people attempting both content programs dropped from 12 to five (although all 12 remained in math, leaving the cohort at 24), and most of those five who remained indicated regretting that decision as it minimized possible learning within all courses. Program developers should never have allowed this choice because providing this option suggested there was a realistic probability of success. Finally, the program only suggested to participants how to complete their preparation for the Exam Option and when to take the exams while in the course sequence.

This pathway should have been more formally stated, and perhaps even required.

A Final Thought

There is currently little research on the necessary conditions to make expedited certification programs as effective as traditional credential programs (Allen, 2003; Rosenberg & Sindelar, 2005). Findings from this case study highlight the importance of structuring programs to meet both academic and affective needs of the intended learners. Addressing the equivalent of a major in mathematics in merely three courses was not nearly sufficient given the content background of the participants. Although the courses expanded participants' content knowledge, the fast pace overwhelmed them, leaving instructors to strategize ways to alleviate anxiety. It is not surprising that most were not able to pass the state exams at first attempt.

Alternative route programs must balance the degree of acceleration with what is needed for success. Beginning an expedited program without adequate content preparation does not realistically facilitate the goals of immediate certification. Additionally, participants in such a program need support to devote the time and energy needed to adequately master a new content area. Participants in this program were not provided any release time from job duties (e.g., classes or meetings), and hence completed all work for the program above and beyond their normal teaching loads. Ideally, developers of the program should be in close communication with districts and school sites to negotiate participants' release from duties outside of teaching.

Alternative programs should provide long-term support once teachers move into new content areas. Insufficient time to process new content understandings and develop pedagogical content knowledge often translates into less than ideal pedagogical choices (Philipp, 2007). This program was content driven and focused on participants passing certification exams; the design did not build in extended support in classroom settings. Mentoring has been a recommended component for both teacher effectiveness and efficacy (Ross, 1995); however, few alternatively structured certification programs have a focus on mentoring (Humphrey & Wechsler, 2007). An additional professional development component that extends into the secondary classroom would help teachers to translate new content knowledge into effective choices in practice.

Furthermore, this program targeted teachers changing disciplines, as opposed to non-educators changing careers; all participants had existing certifications. Most alternative programs are focused on initial certification; with fewer teaching jobs opening overall, and more jobs available in science and mathematics, there will likely be increased need for this type of alternative program. Given the current situation, one area needing further investigation is whether or not existing knowledge of pedagogy translates to the teaching of newly learned content. Research-based models that facilitate expedited pathways preparing teachers for successful secondary mathematics and science teaching, as well as enhancing the understanding of effective, creative instruction in teacher education, must be created.

References

Allen, M. (2003). *Eight questions on teacher preparation: What does the research say? A summary of the findings.* Retrieved from http://ecs.org/html/educationIssues/teachingquality/tpreport/home/summary.pdf

Ball, D., Lubienski, S., & Mewborn, D. (2001). Research on teaching: The unsolved problem of teachers' mathematical knowledge. In V. Richardson (Ed.), *Handbook of research on teaching* (4th ed., pp. 433–456). New York: Macmillan.

Benken, B. M., & Brown, N. (Fall, 2008). Moving beyond the barriers: A re-defined, multi-leveled partnership approach to mathematics teacher education. *Issues in Teacher Education, 17*, 63–82.

California Subject Examinations for Teachers. (2009). Retrieved from www.cset.nesinc.com.

Chin, E., & Young, J. W. (2007). A person-oriented approach to characterizing beginning teachers in alternative certification programs. *Educational Researcher, 36*, 74–83.

Farmer, J. D., Gerretson, H., & Lassak, M. (2003). What teachers take from professional development: Cases and implications. *Journal of Mathematics Teacher Education, 6*, 331–360.

Feistritzer, C. E., Haar, C. K., Hobar, J. H., & Scullion, A. B. (2005). *Alternative teacher certification.* Washington DC: National Center for Education Information.

Hembree, R. (1990). The nature, effects, and relief of mathematics anxiety. *Journal for Research in Mathematics Education, 21,* 33–46.

Hill, H. C., Rowan, B., & Ball, D. L. (2005). Effects of teachers' mathematical knowledge for teaching on student achievement. *American Educational Research Journal, 42*, 371–406.

Humphrey, D. C., & Wechsler, M. E. (2007). Insights into alternative certification: Initial findings from a national study. *Teachers College Record, 109*, 483–530.

LaTurner, R. J. (2002). Teachers' academic preparation and commitment to teach math and science. *Teaching and Teacher Education, 18*, 653–663.

National Council of Teachers of Mathematics. (2000). *Principles and standards for school mathematics*. Reston, VA: National Council of Teachers of Mathematics.

National Mathematics Advisory Panel. (2008). *Foundations for success: The final report of the National Mathematics Advisory Panel*. Retrieved from www.ed.gov/about/bdscomm/list/mathpanel/report/final-report.pdf

Philipp, R. A. (2007). Mathematics teachers' beliefs and affect. In F. Lester (Ed.), *Second handbook of research on mathematics teaching and learning* (pp. 257–318). Charlotte, NC: Information Age Publishing.

Rosenberg, M. S., & Sindelar, P. T. (2005). The proliferation of alternative routes to certification in special education: A critical review of the literature. *Journal of Special Education, 39*, 117–127.

Ross, J. A. (1995). Strategies for enhancing teachers' beliefs in their effectiveness: Research on a school improvement hypothesis. *Teachers College Record, 97,* 227–251.

Stake, R. E. (1995). *The art of case study research.* Thousand Oaks: Sage Publications.

United States Department of Education. (2002). *Meeting the highly qualified teachers challenge: The secretary's annual report on teacher quality.* Washington, DC: Office of Postsecondary Education.

United States Department of Education. (2004). *Alternative routes to certification.* Washington, DC: Author.

U. S. Congress. (2001). *No Child Left Behind Act of 2001.* Public Law 107-110. 107th Congress. Washington, DC: Government Printing Office.

Wilson, M., & Cooney, T. J. (2002). Mathematics teacher change and development: The role of beliefs. In G. Leder, E. Pehkonen, & G. Töerner (Eds.). *Beliefs: A hidden variable in mathematics education?* (pp. 127–147). Dordrecht, The Netherlands: Kluwer Academic Publishers.

Zientek, L. R. (2007). Preparing high quality teachers: Views from the classroom. *American Educational Research Journal, 44,* 959–1001.

Babette M. Benken is associate professor and graduate advisor for mathematics education at California State University, Long Beach. Her research interests include models of teacher development, and the role knowledge, beliefs, and context play in shaping practice. Her email address is bbenken@csulb.edu.

Susan Gomez-Zwiep is an assistant professor of science education at California State University, Long Beach. She studies both preservice and inservice teacher development. She can be reached at sgomezwp@csulb.edu

Kosko, K. W., Norton, A., Conn, A., and San Pedro, J. M.
AMTE Monograph 7
Mathematics Teaching: Putting Research into Practice at All Levels
© 2010, pp. 207–223

14

Letter Writing: Providing Preservice Teachers with Experience in Posing Appropriate Mathematical Tasks to High School Students

Karl W. Kosko
Virginia Polytechnic Institute and State University

Anderson Norton
Virginia Polytechnic Institute and State University

Angie Conn
Virginia Polytechnic Institute and State University

Jan Michael San Pedro
Virginia Polytechnic Institute and State University

Letter writing provides preservice teachers with opportunities to pose mathematical tasks to students and gain a deeper understanding of what makes a task cognitively demanding (Crespo, 2003; Boston & Smith, 2009). This article describes the process of implementing a letter-writing project for preservice teachers and shares results concerning the project's impact on a cohort of preservice secondary school teachers. In particular, the project enhanced the pedagogical content knowledge of preservice secondary teachers in terms of their ability to elicit mathematical processes and higher levels of cognitive demand from students. Data collected over the course of a semester is included to support claims and to provide readers information about aspects of implementation that seem to better prepare preservice secondary teachers.

Thompson, Carlson and Silverman (2007, p. 416) wrote, "Tasks affect learners, or not, because the learner accepts what is offered, or not, in the context of his or her own meanings, goals, interests, and commitments. It is in this sense that one must develop tasks with the learner in mind." However, before entering the classroom, preservice secondary school teachers do not always have opportunities to develop mathematical tasks with the learner in mind. In studying both inservice and preservice teachers' problem posing, Silver, Mamona-Downs, Leung, and Kenney (1996) advocated "a need to provide more opportunities for prospective and in-service teachers to engage in mathematical problem posing and to analyze the emerging problems for their feasibility and their quality" (p. 305).

Letter writing has provided one type of successful experience in problem posing and task generating for preservice teachers (e.g., Crespo, 2003; Norton & Rutledge, 2007; Phillips & Crespo, 1996). Letter writing, as defined by Crespo (2003), is similar to a "pen pal" project in which preservice teachers pose mathematical tasks to students and students respond. Thus, it is a useful simulated teaching experience with an element of authenticity. Letter writing also provides additional experience in learning how to communicate mathematically with others (Goodman, 2005).

Engaging preservice secondary teachers in letter writing provides a host of benefits to their development as mathematics teachers. This article relays teacher educator experiences with letter writing and shares practical information for implementing a letter writing project. It begins with a brief background of research on letter writing and the reasoning that led to the authors' implementation of the project. The article then goes on to describe the kinds of growth preservice secondary teachers experienced from their participation in the project.

Background and Implementation

Engaging preservice secondary teachers in letter writing was inspired in part by Crespo's (2003) work, which was in turn inspired by Fennell (1991). Crespo investigated preservice teachers' problem posing in writing letters about mathematics to fourth grade students. Initially, preservice teachers asked for simple answers, but over time they asked for more explanation and justification. Written problems became less straightforward and questions were asked that allowed for better understanding of the students' thinking.

Crespo (2003) noted that while the nature of the mathematics problems changed greatly over the course of letter writing, "… these changes did not happen overnight and were not self generated…left to their own devices preservice teachers' tendencies were to pose unproblematic problems to their pupils" (p. 264). Crespo argued that by having an authentic audience in terms of the student they were writing, preservice teachers were able to adjust their views as to what a properly posed problem looks like.

Building on Crespo's work, the second author engaged previous cohorts of preservice teachers in letter-writing projects. These projects were deemed successful in terms of preservice teachers' growth in posing cognitively demanding tasks and focusing on students' mathematical thinking (Norton & Rutledge, 2007). However, limitations in the design and implementation of those previous projects showed a need for additional classroom support to interpret students' responses and for a longer interval between tasks and responses. Incorporating appropriate changes, the authors engaged a new cohort of preservice teachers in a letter-writing project to support their ability to pose mathematical problems and improve their understanding of students' thinking.

The preservice teachers were enrolled in the first of a sequence of senior-level mathematics courses entitled "Mathematics for Secondary School Teachers." The course links secondary and college-level mathematics by extending secondary school topics in ways that require preservice teachers to explain and justify their thinking. In particular, the fall semester focused on discrete mathematics, including topics typically arising in high school (e.g., permutations and combinations), college (graph theory), or both (matrices). Thirty preservice teachers were enrolled in the class that formed the basis for this study. The class met for 75 minutes every Tuesday and Thursday; Thursdays were dedicated to the letter-writing project.

During the first few weeks, preservice teachers were oriented to the project by considering readings such as sections from Stein, Smith, Henningsen, and Silver (2000), Polya's *How to Solve It* (1957), and *Principles and Standards for School Mathematics* (2000) by the National Council of Teachers of Mathematics (NCTM). The preservice teachers also wrote a letter of introduction (using pseudonyms) describing their interests in mathematics, teaching, and extracurricular activities. These letters were scanned and emailed to a high school teacher who taught precalculus in a Cincinnati suburb. The high school students responded by sharing their interests, including descriptions of what they liked and did not like about mathematics. In the third week of the college course, the preservice teachers designed and posed their first tasks to the high school students, attempting to connect to the students' interests and engage them in a discrete mathematics task. These tasks were embedded in letters that included additional personal communication and encouragement.

The mathematical tasks were expected to embed the different NCTM (2000) process standards of problem solving, reasoning and proof, communication, connections, and representation and to employ progressively higher levels of cognitive demand (Stein et al., 2000). The preservice teachers were guided by an assessment form that encouraged designing tasks that incorporated the process standards and high levels of cognitive demand as well as at least one content standard for discrete math from the state of Virginia (VDOE, 2009). After designing a task, each preservice teacher predicted what processes and levels of cognitive demand the task would elicit.

Tasks and their companion letters were collected on Tuesdays, emailed to the high school on Wednesdays, and distributed to high school students on Thursdays. The high school students were allowed to spend some class time (typically 20 to 30 minutes) working on the tasks, but students often worked on the tasks outside of class as well. The preservice teachers generally received responses the following Tuesday, and the college course allowed workshop time on Thursdays to discuss and analyze responses. The preservice teachers also began to conceptualize new tasks to build on the students' thinking during the Thursday workshops. The cycle continued the next week and was in effect for twelve weeks, producing a total of six tasks and six responses. The preservice teachers then wrote reflective papers evaluating their growth as teachers based on their experience with the letter-writing project.

As found in similar implementations (Norton & Rutledge, 2007), preservice teachers demonstrated improvements over the semester in their ability to engage students in different mathematical processes at a high level of cognitive demand. An example of this growth is provided in Figure 1, which shows an excerpt of one preservice teacher's predictions and assessments for Tasks 1 and 5. The increase in the amount of detail provided from Task 1 to Task 5 is characteristic of the cohort of preservice teachers involved in letter writing. In the following sections, growth among the cohort as a whole is described, followed by a discussion of implications for teacher education programs.

NCTM Process Standard	Predicted? (check box)	Observed?	Evidence
Problem Solving	✓	X	[illegible]
Reasoning & Proof	✓	X	
Communication	✓	✓	She let me know how she solved it & told me that my problem wasn't clearly stated enough.
Connections		X	
Representation	✓	✓	She made a chart to organize her thinking.
Level of Cognitive Demand	**Predicted (M, P, C, or D)**	**Observed**	**Evidence**
	C	P	
Discrete Math VA SOL	**Predicted (DM1, DM2, etc.)**	**Observed**	**Evidence**
	Dm 13	X	No real [illegible]

use complete sentences.

Task 1

NCTM Process Standard	Predicted? (check box)	Observed?	Evidence
Problem Solving			
Reasoning & Proof	✓	✓	She took her answer & reasoned out why it works, then showed examples of it working & not working.
Communication	✓	✓	She clearly stated how & why she got her answer.
Connections	✓	✓	I feel as though she actually has made the connection as to why you need an even number of edges from each vertex.
Representation		✓	She drew a picture to help explain & show her explanation of why Euler circuits need to have an even number of edges from each vertex.
Level of Cognitive Demand	**Predicted (M, P, C, or D)**	**Observed**	**Evidence**
	C	C	She needed to take what she had already learned & figure out why it worked, and [illegible] think she [illegible] [illegible] with the why.
Discrete Math VA SOL	**Predicted (DM1, DM2, etc.)**	**Observed**	**Evidence**
	DM2	DM2	Even though she's working with [illegible], she's thinking about Euler circuits.

Task 5

Figure 1. Comparison of preservice teacher predictions and assessments in Task 1 and Task 5.

Process Standards

As noted earlier, the preservice teachers made predictions and assessments about whether their tasks would engage (or had engaged) students in each of the five NCTM (2000) process standards. Two of the authors also assessed student responses in order to measure the preservice teachers' progress, both in designing better tasks and in assessing student responses. A student response indicated *problem solving* if there was evidence that the student had used a novel approach to a solution. In other words, the student response needed to indicate cognitive struggle and not a simple application of a familiar procedure or formula. *Reasoning and proof* was indicated by evidence of a conjecture and an attempt to prove the conjecture. Evidence of *communication* included describing procedures or solution strategies and/or justification of claims. Simply stating an answer or showing scratch work was not counted as communication; the student writing had to clearly indicate the processing of mathematical ideas. Evidence of making *connections* came from student responses that connected mathematical ideas or linked mathematics to other contexts. Finally, *representation* was attributed if unsolicited visual representations were used as part of the process of resolving the task.

As part of the assignment, the preservice teachers had to demonstrate that their high school student letter-writing partners engaged in each process at least once while completing the six tasks. Figure 2 shows the level of elicitation of each process over the semester.

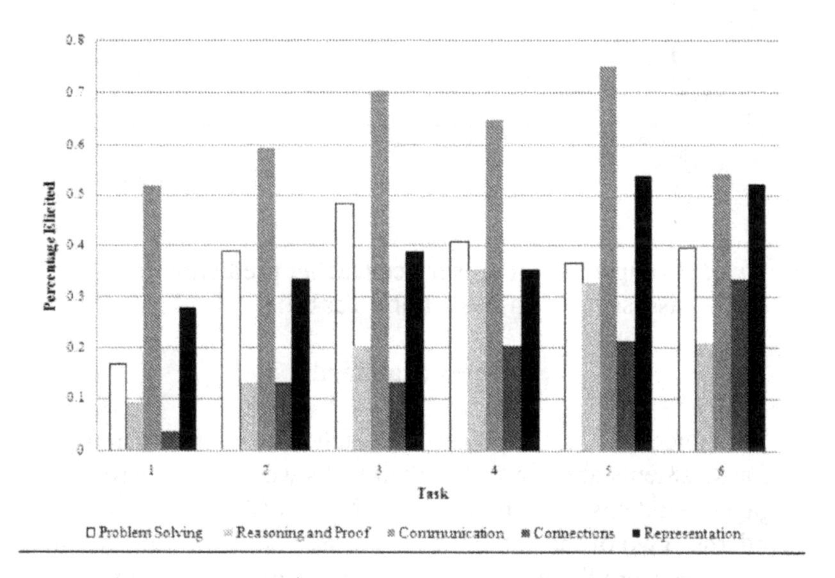

Figure 2. Percent of student elicitation of each process standard by task.

The preservice teachers' improvement in emphasizing mathematical processes was indicated throughout the semester by the quality of their class discussions, by a notable change in their predictions and assessments, and by their comments at the end of the course. For example, one preservice teacher stated, "I didn't understand problem solving at the very beginning, but after talking to [the instructor], I realized that in order to engage my student in problem solving, I had to give her something that she would struggle with." By "struggle" the preservice teacher meant that the student needed to be challenged to think critically.

By examining the preservice teachers' predictions and assessments, two of the authors found that preservice teachers had high expectations of the high school students' responses during the first week of letter writing. Over fifty percent of the preservice teachers predicted the elicitation of every process standard except for connections, but the authors found a much lower percentage elicited from students. Over the course of the letter-writing project, preservice teachers' assessments of high school students' responses became more closely aligned to those of the authors. This convergence signifies a better understanding of the practical meaning of the process standards.

Levels of Cognitive Demand

In addition to examining the mathematical processes demonstrated through students' responses, the preservice teachers were asked to predict and assess cognitive demand for the tasks they created. Specifically, they used the four levels of cognitive demand defined by Stein et al. (2000): memorization; procedures without connections; procedures with connections; and doing mathematics. *Memorization* is indicated when a student provides an answer or statement from memory or outside resources such as a textbook. The cognitive level of *procedures without connections* is evidenced when a student uses a provided algorithm to solve for an answer. A student exhibiting *procedures with connections* may not use a provided algorithm but instead investigates the mathematical task through a solution strategy of his or her own making. *Doing mathematics* entails the same indicators as procedures with connections, but also includes conjecturing on the part of the student. When students are doing mathematics, they are not simply creating a solution strategy but testing and justifying it (Stein & Smith, 1998; Stein et al., 2000).

Literature on cognitive demand reveals that preservice teachers have difficulty distinguishing between the two highest levels of cognitive demand: procedures with connections and doing mathematics (Osana, Lacroix, Tucker, & Desrosiers, 2006). This was also seen in the current study. Even in later weeks of the letter-writing project, many preservice teachers had difficulty discerning doing mathematics from procedures with connections. However, preservice teachers' understanding of the different levels of cognitive demand did increase overall. In evaluating the preservice teachers' predictions and assessments regarding cognitive demand, it was found that both converged with the authors' assessments.

In addition to increasing their understanding of different levels of cognitive demand, the preservice teachers generally improved their ability to elicit higher levels of cognitive demand over the course of the semester (see Figure 3). Most remarkable about the preservice teachers' progress is the fact that their tendency to emphasize lower levels of cognitive demand (*m* for memorization and *p* for procedures without connections) decreased from approximately 90% in Week 1 to around 40% in Week 6. This difference is illustrated by the line superimposed on Figure 3. There is a corresponding trend of increasing levels of cognitive demand (*c* for procedures with connections and *d* for doing mathematics). The two higher levels increased from less than 10% of students in the first task to almost 60% of students in the sixth task.

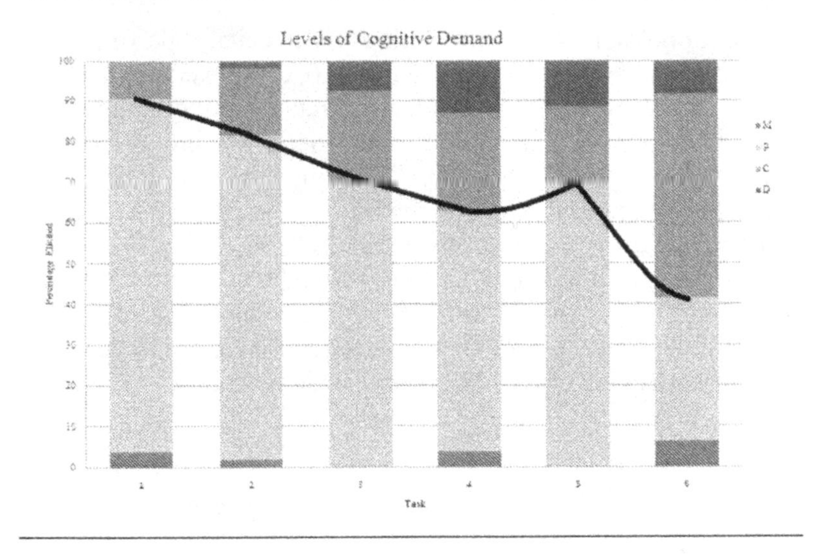

Figure 3. Levels of cognitive demand elicited by task.

Overall, these findings are similar to those of Arbaugh and Brown (2005), who found that engaging teachers in assessing tasks for their level of cognitive demand increased the teachers' pedagogical knowledge and their use of high-level tasks. One preservice teacher in this study commented as follows:

> I had to recognize what my student was good at, and how good she was at mathematics. Once I figured out what her level was, I was able to create problems that challenged her enough where she had to think and work a little bit harder than she was used to. If the problem was too easy she would be able to complete the problem without really thinking.

Several other preservice teachers exhibited this awareness of the importance of knowing students' capabilities in order to challenge them. We conjecture that such statements not only imply familiarity with levels of cognitive demand, but also indicate increased pedagogical knowledge in general. This increase is evident in the student responses shown in Figure 4. In Task 2, a high school student partner was asked to find the number of possible song combinations for a recital if performing three songs from a list of 10 songs. Task 6, completed by the same student, focused on forming a continuous ring of dominoes with sets of dominos ranging from zero to nine or zero to six. For both sets, each domino included all combinations of numbers within the set (e.g., a domino in the first set may have the numbers 0 and 0 or 3 and 6, denoted as (0,0) and (3,6) respectively). A continuous ring was formed when the dominoes could be matched (3s to 3s, 4s to 4s, etc.) in such a way as to form a circle with the dominoes. As seen in Figure 4, this task proved challenging to the high school student.

Excerpt from Task 2:

I think I understand combinations and permutations now, thanks. I was doing a permutation earlier, and I'll use one now. To choose 3 songs for my repertoire, I should be able to multiply 10 × 9 × 8 to get my answer. That would give me 720 song combinations.

Excerpt from Task 6:

For a set with 0–6 . . .

(0,0) → (0,1) → (1,2) → (2,3) → (3,4) → (4,0) → (0,6) → (6,1) → (1,3) → (3,5) → (4,6) → (6,2) → (2,5) → (5,6) → (6,3) →
(3,0)(0,2) → (2,4) → (4,3) → (3,1) → (1,6) → (6,0) → (0,3) → (3,6) → (6,4) → (4,1) → (1,5) → (5,0) → (0,4) → (4,4) →
(4,2) → (2,1) → (

I could not link this together . . .

Hopefully Set 6 is the one that does not work and the deluxe does. I'm not sure how else to do this.

Let me try a vertex edge graph for the dominoes.

So actually, the set of 6 works!

For the deluxe set . . . I ended at 5 which does not connect back to 0. So the deluxe set does not work.

As for the game, whoever goes first seems to win. This is because there are 10 lines you can have without forming a triangle. I'm not sure about the students in the room.

Figure 4. Evidence of levels of cognitive demand in early and later student work.

Note that in Task 2, the student demonstrated a relatively low level of cognitive demand as indicated by the use of a provided algorithm. Over the semester, the preservice teacher who worked with this student refined efforts to challenge the student to think differently about mathematics. In Task 6, there is a notable difference in the amount of detail provided not only about the procedures used but also about the student's thought processes. Additionally, the tone of the student's writing has changed. The later entry replaces the use of an algorithm with conjecture and justification through use of a representation. The student's statement "So actually, the set of 6 works!" conflicts with the student's prediction, but shows how she used the vertex solution strategy to further her understanding of the topic.

While the general trend of growth in the letter-writing project was positive (see Figure 3), significant shifts in student performance like those shown in Figure 4 were not evident in every preservice teacher/student relationship. It takes time and experience for preservice teachers to learn how to elicit higher levels of cognitive demand and different mathematical processes from students. Letter writing allows preservice teachers to have such experiences before they enter the classroom.

Discussion and Future Directions

Mathematics teachers often believe their students are mathematically engaged at a more sophisticated level than the students actually are (Carpenter, Fennema, Peterson, & Carey, 1988). The use of writing in mathematics helps correct such misconceptions as teachers gain a deeper knowledge of their students' mathematics (Moskal, 1997). Engaging a cohort of preservice teachers in letter writing improved their ability to predict levels of cognitive demand and to implement the NCTM (2000) process standards in ways that elicit mathematical processes from students. Likewise, requiring preservice teachers to focus on mathematical processes and levels of cognitive demand likely improved their pedagogical knowledge and their ability to pose mathematical tasks. This improvement was not only associated with the preservice teachers' increased understanding of cognitive demand and mathematical processes, but was also related to their knowledge of particular students, both personally and pedagogically. The preservice teachers gained an appreciation for students' mathematical thinking as they designed and posed mathematical tasks with particular students in mind.

As with any implementation in teaching, further adjustments can and should be considered. The findings discussed here are from the third implementation of the letter-writing project, which has evolved based on findings from previous implementations. One of the projects' greatest strengths is its flexibility, allowing for the modification of resources, workshop activities, rubrics, and grade levels, while continuing to provide preservice teachers with access to students. The authors encourage other educators to make modifications in further implementations of the letter-writing project. One such modification could engage students in the use of specific software (e.g., Microsoft Word's Equation Editor or Geometer's Sketchpad) or other forms of technology in writing about mathematics.

References

Arbaugh, F., & Brown, C. (2005). Analyzing mathematical tasks: A catalyst for change? *Journal of Mathematics Teacher Education, 8*, 499–536.

Boston, M. D., & Smith, M. S. (2009). Transforming secondary mathematics teaching: Increasing the cognitive demand of instructional tasks used in teachers' classrooms. *Journal for Research in Mathematics Education, 40*, 119–156.

Carpenter, T. P., Fennema, E., Peterson, P. L., & Carey, D. A. (1988). Teachers' pedagogical content knowledge of students' problem solving in elementary arithmetic. *Journal for Research in Mathematics Education, 19*, 385–401.

Crespo, S. (2003). Learning to pose mathematical problems: Exploring changes in preservice teachers' practices. *Educational Studies in Mathematics, 52*, 243–270.

Fennell, F. (1991). Diagnostic teaching, writing, and mathematics. *Focus on Learning Problems in Mathematics, 13*, 39–50.

Goodman, R. E. (2005). Using letter-writing to enhance a calculus course. *PRIMUS: Problems, Resources, and Issues in Mathematics Undergraduate Studies, 15*, 298 – 302.

Moskal, B. (1997, March). Open-ended mathematics tasks: How did a middle school teacher interpret and use information acquired through the examination of student responses? Paper presented at the Annual meeting of the American Educational Research Association. Chicago, IL.

National Council of Teachers of Mathematics (2000). *Principles and standards for school mathematics*. Reston, VA: NCTM.

Norton, A., & Rutledge, Z. (2007). Assessing cognitive activity in the context of letter writing. In T. Lamberg & L. Wiest (Eds.), *Proceedings of the Twenty-Ninth Annual Meeting of the North American Chapter of the International Group for the Psychology of Mathematics Education* (pp. 159–167). Stateline (Lake Tahoe), NV: University of Nevada, Reno.

Polya, G. (1957). *How to solve it* (2nd ed). New York: Doubleday Anchor Books.

Osana, H., Lacroix, G., Tucker, B., & Desrosiers, C. (2006). The role of content knowledge and problem features on preservice teachers' appraisal of elementary mathematics tasks. *Journal of Mathematics Teacher Education, 9*, 347–380.

Phillips, E., & Crespo, S. (1996). Developing written communication in mathematics through math pen pal letters. *For the Learning of Mathematics, 16*, 15–22.

Silver, E. A., Mamona-Downs, J., Leung, S. S., & Kenney, P. A. (1996). Posing mathematical problems: An exploratory study. *Journal for Research in Mathematics Education, 27*, 293–309.

Stein, M. K., & Smith, M. S. (1998). Mathematical tasks as a framework for reflection: From research to practice. *Mathematics Teaching in the Middle School, 3*, 268–275.

Stein, M. K., Smith, M. S., Henningsen, M. A., & Silver, E. A. (2000). *Implementing standards-based mathematics instruction: A casebook for professional development*. New York: Teachers College, Columbia University.

Thompson, P. W., Carlson, M. P., & Silverman, J. (2007). The design of tasks in support of teachers' development of coherent mathematical meanings. *Journal of Mathematics Teacher Education, 10*, 415–432.

Virginia Department of Education [VDOE]. (2009). Mathematics standards of learning for Virginia public schools. Retrieved from http://www.doe.virginia.gov/testing/sol/standards_docs/mathematics/index.shtml

---

Appendix

Letter Writing Project
Task Analysis Form

Name ____________________
Date __________________

NCTM Process Standard	**Predicted? (check box)**	**Observed?**	**Evidence for Observation, or Reasons Why Expected Processes Were not Observed**
Problem Solving			
Reasoning & Proof			
Communication			
Connections			
Representation			
Level of Cognitive Demand	**Predicted (M, P, C, or D)**	**Observed**	**Evidence**

Discrete Math VA SOL	Predicted (DM1, DM2, etc.)	Observed	Evidence

Use the student response to your task to reflect on and analyze ways in which you can improve future task design.

Karl Kosko is a post-doctoral fellow at the University of Michigan and recent graduate of Virginia Tech. His research interests include factors of mathematical communication, including the self-regulation of mathematical discussion.

Anderson Norton is an assistant professor in the Mathematics Department at Virginia Tech. He teaches content courses for future secondary school teachers and researches how students learn, especially with regard to conjecturing activity.

Angie Conn is a graduate doctoral student at Virginia Tech working in curriculum and instruction for mathematics education. Her email address is aconn@vt.edu

Jan Michael San Pedro is a recent graduate of Virginia Tech, where he was enrolled as a preservice teacher. He currently works in San Diego.

Leatham, K. R., and Peterson, B. E.
AMTE Monograph 7
Mathematics Teaching: Putting Research into Practice at All Levels

15

Purposefully Designing Student Teaching to Focus on Students' Mathematical Thinking

Keith R. Leatham
Brigham Young University

Blake E. Peterson
Brigham Young University

Traditional student teaching often focuses more on classroom management than on how to craft and carry out lessons in ways that engage students in meaningful mathematical activity. This chapter describes five common problems with traditional student teaching: lackluster outcomes, focus on survival over technique, focus on self, isolation, and lack of direction. It then describes a purposeful redesign of the student teaching experience, and how redesigning the structure can help mathematics teacher educators address these common problems.

The efficacy of student teaching is a concern because much of that experience seems to be centered around classroom management and not on how to craft and carry out a lesson in a way that would engage students in meaningful mathematical activity. Although learning to manage student behavior and to survive may be the "reality" of the classroom, one could also view these techniques as peripheral to the central "reality" of teaching, that of facilitating students' learning of mathematics. The question "Why couldn't a focus on students' mathematical thinking and the challenging work of anticipating, eliciting, using and extending that thinking be considered the 'reality' of the classroom?" motivated the authors to examine the student teaching program at Brigham Young University (BYU). At that time, the program left most of the day-to-day structure of student teaching up to the cooperating teacher. Therefore, the authors examined how the cooperating teachers viewed the purpose of student teaching (Leatham & Peterson, 2010).

The typical cooperating teacher felt the purpose of student teaching was for student teachers to learn how to manage real classrooms by interacting with experienced teachers in their classrooms. Eight of the 45 cooperating teachers who participated in the study indicated that the primary purpose of student teaching was to learn about classroom management, with 10 others indicating that classroom management was one of the primary purposes. The cooperating teachers referred a great deal to experience with real teachers, with real classrooms and with real behavior problems; only one cooperating teacher referred to the value of experience with real students' thinking.

One finding of this initial study was that the current BYU student teaching program was not perceived by cooperating teachers as having the purpose of learning to craft and carry out mathematics lessons that effectively anticipated, elicited, and used students' mathematical thinking. This paper discusses how the authors' personal experiences, as well as their understanding of the research literature on student teaching, influenced a purposeful redesign of the student teaching experience toward that purpose.

The Traditional Structure

Student teaching at BYU had followed a traditional model. Student teachers worked with one secondary (grades 6–12) mathematics teacher for 15 weeks. They observed their cooperating teachers during the first few days of the experience, then, in a decision made jointly by the cooperating and student teacher, progressively took over responsibility for the cooperating teacher's classes. The cooperating teacher may have remained as a clear presence in the classroom for the entire 15-week student teaching experience, or may have quickly turned over full control of the classroom to the student teacher. The cooperating teacher may or may not have held regularly scheduled planning meetings, observations, or debriefing sessions with the student teacher. The transition to full control, the presence of the cooperating teacher in the classroom, and the frequency of observations all depended on the style and assumed role of the cooperating teacher. In addition, student teachers were observed approximately once a week by a university supervisor, who then met with the student teacher to discuss the observed lesson. This student teaching structure was very much in keeping with what is described in the literature as a "traditional, apprentice-type program" (McIntyre, Byrd, & Foxx, 1996, p. 173). Although teacher educators often work toward improving this structure (Cochran-Smith, 1991; Zeichner, 2002), it appears to be as common today as it was 30 years ago (Rodgers & Keil, 2007).

Problems with the Traditional Student Teaching Structure

The research community has long documented the problematic nature of the traditional student teaching experience (Brouwer & Korthagen, 2005; Feiman-Nemser & Buchmann, 1985). Based on an analysis of the literature on student teaching and the authors' experiences, five problems with student teaching and ways in which the traditional structure aids and abets those problems are discussed here.

Lackluster Outcomes

One long-standing problem with student teaching is that the experience tends not to accomplish what teacher education programs wish that it would (Cochran-Smith, 1991; Wilson, Floden, & Ferrini-Mundy, 2002; Zeichner, 1981). The traditional structure of student teaching contributes to this problem by lacking clearly delineated learning goals. Because universities seldom relay meaningful expectations with respect to the purpose of student teaching, cooperating teachers and student teachers are left to determine this purpose for themselves, and may conclude the purpose of student teaching is to "experience a real classroom" (Leatham & Peterson, 2010). In the absence of clearly articulated goals, student teaching often falls into the trap of experience for experience sake (Feiman-Nemser & Buchmann, 1985). This lack of explicitly delineated purposes contributed to the authors' dissatisfaction with the nature of the BYU student teaching structure.

Survival over Technique

As mentioned previously, research has found that student teaching tends to focus on survival and classroom management rather than on the craft of teaching. The traditional student teaching structure contributes to this focus because student teachers quickly become full-time teachers by taking over the entire load of the cooperating teacher, including teaching all classes, preparing lessons for those classes, grading, etc. Under such a structure, when the reality of the classroom is viewed as running a classroom more than understanding and facilitating student thinking, it is easy to see how student teaching could be viewed as a time to learn how to survive in the classroom. A common survival tactic is to take minimal ownership of lesson materials and to present lessons that rely on a transmission model for teaching, requiring only that students be quiet and pay attention. The success of such lessons requires a great deal of classroom management; students tend not to behave, as these lessons are far from motivating or engaging.

Focus on Self

Student teachers tend to focus on their teaching to the exclusion of students' learning. Much of the understanding of this phenomenon comes from research on the stages of development of novice teachers (e.g., Fuller, 1969; Maynard & Furlong, 1993). Such research has shown that student teachers work through stages of concern, initially struggling to see themselves as teachers, then worrying about their preparation and knowledge, followed by concerns related to their own performance as a teacher. In the final stage novice teachers express concern for their students. Interpretations of these developmental theories, however, have tended to confuse "description with prescription" (Feiman-Nemser, 2001, p. 24). Although it is valuable to recognize the primary concerns of novice teachers, these concerns should not dictate the focus or structure of teacher education. In fact, research on student teaching seems to take the structure of student teaching for granted, as if understanding what *is* happening in student teaching programs were equivalent to knowing what *could* or *should* happen. Perhaps the structure of the teacher education programs in which developmental studies were conducted was taken for granted and influenced the outcomes of the studies more than the researchers realized. The structure of student teaching may have contributed to the ways the student teachers experienced their concerns. When someone is thrown into the water and is just learning how to swim, it makes sense for them to be more worried about self-preservation than proper stroke technique, let alone the status of other individuals in the pool.

Another aspect of focusing on self is the tendency of novice teachers to assume that their students will think about and learn mathematics the same way they did. Student teachers struggle to see mathematics from their students' perspectives. Teachers need to develop the "capacity to adopt the other's perspective" (Arcavi & Isoda, 2007, p. 114), which is closely tied to the Piagetian notion of decentering. One problem with the traditional structure of student teaching is that it encourages a focus on student teachers' own actions as teachers and thus makes it difficult for them to decenter in order to focus on their students.

Isolation

Teaching in the United States has tended to be isolationist (Labaree, 2000), in that teachers are given a great deal of independence in running their classrooms, are often left alone to do so, and are given very little time to converse or collaborate with fellow teachers. Traditional student teaching has often followed suit. Although student teachers usually experience a degree of collaboration with their cooperating teachers, they are often given a great deal of independence in running their classrooms and are frequently left alone to "see if they can handle it." It is somewhat ironic that a major aspect of student teachers' socialization into schools is learning to perform the job of a teacher in isolation from others.

A Class with No Teacher

Cooperating teachers tend to see themselves more as an experienced colleague than as a teacher educator (Little, 1990; Wang & Odell, 2007), yet the traditional structure of student teaching leaves most of the "curriculum" of the student teaching "class" up to the cooperating teacher. The problem then is that teacher education programs require student teachers to enroll in this class, which has no well-defined curriculum. It is also unclear whether the teacher of this course is the cooperating teacher or the university supervisor. The traditional structure of student teaching contributes to this problem through its lack of clearly defined purposes, as well as by failing to support cooperating teachers in understanding what is expected of them. Similarly, the traditional structure works against the teaching role of the university supervisor, as supervisor roles tend to be more evaluative and their involvement periodic and removed.

A Redesigned Student Teaching Structure

The authors set out to address these problems by redesigning student teaching as they would design any other class. Learning goals for the experience were delineated, and then the structure and the learning activities were designed to meet those goals. As the authors discussed the structure and purpose of the BYU student teaching program, they quickly realized that there was really nothing in place that articulated to cooperating teachers, student teachers or university supervisors what this structure or purpose might be. In addition, they had never actually attempted to articulate that purpose, or to question the degree to which the structure supported that purpose. The authors felt empowered as they articulated their most important purposes for student teaching and envisioned a structure that might support rather than subvert those purposes.

The main purpose for student teaching became a focus on the craft of teaching, where teaching is what one does to facilitate learning. The authors view the craft of teaching as the ability to design lessons that involve important mathematical ideas, to design tasks that help students access those ideas, and then to successfully teach the lesson, which entails effectively launching the lesson, facilitating student engagement with the task, orchestrating meaningful mathematical discussions, and helping to make explicit the mathematical understanding students are constructing (Peterson & Leatham, 2009; Stein, Engle, Smith, & Hughes, 2008). Designing tasks that are accessible to students requires consideration of how students think mathematically. Similarly, to facilitate meaningful student engagement, one must consider how students' previous mathematical experiences come to bear in the current task. Finally, using and building upon student mathematical thinking is a hallmark of orchestrating mathematical discussions. Because of this common thread of students' mathematical thinking, a refined purpose of student teaching emerged: *The primary purpose of student teaching is to learn how to anticipate, elicit and use students' mathematical thinking.* Having determined this purpose, the authors next considered how the current student teaching structure (or lack thereof) facilitated or hindered this purpose as well as how different structures might better facilitate it.

Peterson (2005) spent time in Japan observing the student teaching process at several universities, and noted that the Japanese structure appeared to be in line with the authors' desired purpose of student teaching. The authors considered to what extent the BYU structure could adopt features of the Japanese structure in order to address the identified problems with traditional student teaching. The possibilities from Japanese student teaching and the authors' explicitly articulated purpose for student teaching guided the redesign of the structure of student teaching.

Student Teaching Pairs and Clusters

One key aspect of the new structure placed student teachers in pairs with one cooperating teacher. Student teachers were encouraged to plan all lessons together with the cooperating teacher, but the lessons were taught by one student teacher at a time with the peer acting as observer. To facilitate this setup, student teachers taught the same classes during different periods of the day. For example, if the cooperating teacher taught four algebra classes and two geometry classes, each student teacher would teach two algebra classes and one geometry class, alternating who taught each subject first. They taught these same class periods for the duration of student teaching. Either two or three pairs of student teachers were placed in neighboring schools and organized into a cluster. This model allowed student teachers to observe each other several times during the first weeks of student teaching.

Learning-to-Teach Activities

During the first five weeks of student teaching, the student teachers taught only three times but were given many "learning to teach" activities. After the first five weeks, these activities were replaced by regular daily teaching of half of the cooperating teacher's class load. The student teacher pairs continued to prepare their lessons together in consultation with the cooperating teacher and to observe each other teaching the lessons. Descriptions of the learning-to-teach activities for the first five weeks follow.

Daily journals. Each day student teachers identified an issue they wanted to learn more about that day (e.g., launching a lesson, orchestrating class discussion, managing the classroom, establishing classroom norms). They then briefly summarized what they did during each class period of the day, accounting for the way they spent their time (given that they were only teaching one lesson a week during this portion of the experience). Finally they described some interesting student mathematical thinking that they observed that day. In describing that thinking, they also discussed why it was interesting, what questions they might ask the student if they were teaching, and how they might use the thinking in a class discussion.

Focused observations. During each of the first five weeks, student teachers were expected to complete two focused observations, conducted in a variety of different classes. The foci for each of the first five weeks were as follows: the flow of the lesson, classroom discourse, individual student mathematical experiences, questions and answers, and teachable moments.

These observations were synthesized in a brief paper (two to three pages).

Student interviews. Each student teacher conducted three student interviews, choosing students who had provided interesting student thinking on an exam, quiz, or homework assignment. The student teacher interviewed the student about the mathematical thinking behind their written work and then synthesized the results of the interview in a brief paper.

Teach/Observe/Reflect cycle. During the first two weeks student teachers did not teach any lessons and only completed the learning-to-teach activities. In the second week, each cooperating teacher in a cluster taught lessons that were observed by student teachers in that cluster. During the lesson student teachers and the university supervisor moved around the classroom taking notes on students' mathematical thinking, similar to the "public lesson" phase in lesson study (Lewis, 2002). Later that day during a preparation period or after school, the cooperating teacher and all observers held a formal reflection meeting. The university supervisor facilitated the meeting and began by asking the cooperating teacher to answer three questions: (a) What was the goal of your lesson? (b) How was your lesson designed to meet that goal? and (c) How do you feel the lesson played out? The student teachers were then encouraged to ask the cooperating teacher questions and given an opportunity to make comments about the lesson. Finally, the university supervisor offered concluding comments about the lesson.

In Weeks 3–5 each student teacher taught one lesson per week. The pairs of student teachers coordinated with the cooperating teacher to determine the content and worked together to plan the lesson. They each taught the same lesson during different periods on the same day. All other members of the cluster, the cooperating teacher, and the university supervisor observed the lessons. The subsequent reflection meeting focused on both lessons. A student teacher who had not taught that day acted as facilitator. Thus the initial 20–45 minutes of the reflection meeting was a conversation among student teachers. The cooperating teacher and university supervisor reserved their comments for the conclusion of the reflection meeting, focusing on big principles of teaching mathematics illustrated in the observed lesson and in the foregoing student teacher discussion. One final Teach/Observe/Reflect cycle took place in Week 14.

Reflection papers. At the conclusion of each week's Teach/Observe/Reflect cycle, student teachers wrote a longer reflection paper (five to six pages) on varying topics: Week 2—Reflect on the two or three lessons taught by the cooperating teachers; Week 3—Reflect on the lessons taught by you and your partner student teacher; Week 4—Reflect on your observations of the other student teachers; Week 5—Reflect on all of the lessons taught and observed so far; and Week 14—Reflect on your own lesson. The student teachers were instructed to articulate several pedagogical principles learned over the past 13 weeks and use them as a reflective lens.

Structurally Addressing the Problems with Student Teaching

This section returns to the five identified problems with traditional student teaching and discusses how the authors' revised student teaching structure addressed those problems.

Lackluster Outcomes

It is easy to attribute lackluster student teaching outcomes to the schools and classrooms in which student teaching takes place. It is possible, however, that the structure of student teaching and its associated implicit or explicit purpose may contribute greatly to this problem. In other words, the structure itself may "teach" unwanted lessons on teaching. To address this problem, the purpose of student teaching was clearly articulated with the structure designed to accomplish that purpose. In addition, learning-to-teach activities were designed and implemented. Through structural changes related to specific learning goals, the university took greater responsibility for the outcomes of student teaching.

Survival over Technique

Because a *de facto* focus on survival was one of the main problems addressed in the restructuring, many aspects of the new structure helped to solve this problem while fixing the focus on the craft of teaching. One major structural change was the placement of two student teachers with one cooperating teacher. This pairing facilitates more time spent planning for instruction and provides mutual support. In addition, advantages of partnership teaching contribute directly to the newly articulated purpose. "If to learn to teach is to learn to manage…then partnership teaching has an obvious disadvantage. However, if student teaching's primary purpose is to…expand one's knowledge of methods and of children…then partnership teaching has an advantage" (Bullough, et al., 2003, p. 71).

The learning-to-teach activities demonstrate that the emphasis of this new structure of student teaching was on learning about the craft of teaching and not on peripheral aspects of the job of teaching. In addition, having the student teachers spend a week planning each of the first three lessons communicated the message that they were to learn about the craft of teaching before they began to learn about running the classroom. Student teachers spent the first five weeks focusing on students' mathematical thinking and on crafting lessons related to that thinking. In addition, during the final 10 weeks of student teaching, student teachers spent half of each day observing a peer teach the jointly planned lesson, giving them regular opportunities to see how lessons played out with real students without having the pressure of being in charge.

The structure helped to minimize a focus on survival in one other important way—students are less likely to misbehave with several adults walking around the classroom observing their mathematical thinking. Although the student teachers felt the added pressure of being observed, there were few problems with classroom management during the lessons and the post lesson discussions naturally centered on the lesson and on students' mathematical thinking.

Focus on Self

The learning-to-teach activities were designed for the purpose of learning how to anticipate, elicit and use students' mathematical thinking. Thus, student teachers were repeatedly asked to think about mathematics from the perspective of the students. These activities helped to decenter the student teachers and encouraged them to adopt their students' perspective (Arcavi & Isoda, 2007). In particular, the daily activity of describing and analyzing interesting student thinking in daily journals opened their eyes to students' mathematics. Similarly, when observing student thinking during their peers' lessons (without interacting with the students), student teachers were able to see students struggle with mathematics and attempt to resolve problems on their own.

Isolation

Opportunities to step away from the typical isolation of a teacher occurred on many levels. First, because student teachers were expected to plan their lessons in pairs, they collaborated on a daily basis the entire semester. Secondly, student teachers used their time during the first five weeks observing the cooperating teacher and other teachers in the school as well as teachers in the other cluster schools. Finally, the Teach/Observe/Reflect cycle placed student teachers in the context of observing a peer, having a conversation with peers about the craft of teaching, and then writing a reflection paper based, in part, on these interactions.

A Class with No Teacher

There are several ways the revised structure helped cooperating teachers to be teacher educators—to consider the student teachers as their students. First, the cooperating teachers were asked to do so. Before the beginning of each semester, the authors met with university supervisors and cooperating teachers to discuss the student teaching structure and logistics of scheduling. The main purposes of the student teaching experience and the role of the cooperating teacher and university supervisor in accomplishing those purposes were discussed. Second, a critical part of each reflection meeting was the concluding discussion led by the cooperating teacher. Because the student teachers were encouraged to have a conversation among themselves for the first part of the meeting, it was the cooperating teacher's responsibility to listen to the main points discussed and take on the role of a teacher educator in the concluding discussion. They were expected to build on student teachers' thinking in order to focus on big principles of teaching—thus teaching student teachers in the same manner the student teachers were learning to teach their students. Finally, pairing student teachers with one cooperating teacher halved the number of required cooperating teachers and allowed better placements.

Conclusion

The results of this experiment in purposefully redesigning the structure and focus of student teaching are promising. Dramatic changes took place in the nature of student teacher conversations and in their ability to focus on the craft of teaching. The authors hope this account of a purposeful redesign of student teaching is beneficial to others who seek to make student teaching a capstone university experience in learning to teach.

References

Arcavi, A., & Isoda, M. (2007). Learning to listen: From historical sources to classroom practice. *Educational Studies in Mathematics, 66*, 111–129.

Brouwer, N., & Korthagen, F. (2005). Can teacher education make a difference? *American Educational Research Journal, 42*, 153–224.

Bullough, R. V., Jr., Young, J., Birrell, J. R., Clark, D. C., Egan, M. W., Erickson, L., et al. (2003). Teaching with a peer: A comparison of two models of student teaching. *Teaching and Teacher Education, 19*, 57–73.

Cochran-Smith, M. (1991). Reinventing student teaching. *Journal of Teacher Education, 42*, 104–118.

Feiman-Nemser, S. (2001). Helping novices learn to teach: Lessons from an exemplary support teacher. *Journal of Teacher Education, 52*, 17–30.

Feiman-Nemser, S., & Buchmann, M. (1985). Pitfalls of experience in teacher preparation. *Teachers College Record, 87*, 53–65.

Fuller, F. (1969). Concerns of teachers: A developmental conceptualization. *American Educational Research Journal, 6*, 207–226.

Labaree, D. F. (2000). On the nature of teaching and teacher education: Difficult practices that look easy. *Journal of Teacher Education, 51*, 228–233.

Leatham, K. R., & Peterson, B. E. (2010). Secondary mathematics cooperating teachers' perceptions of the purpose of student teaching. *Journal of Mathematics Teacher Education, 13,* 99–119.

Lewis, C. C. (2002). *Lesson study: A handbook for teacher-led instructional change*. Philadelphia, PA: Research for Better Schools.

Little, J. W. (1990). The mentor phenomenon and the social organization of teaching. *Review of Research in Education, 16*, 297–351.

Maynard, T., & Furlong, J. (1993). Learning to teach and models of mentoring. In D. McIntyre, H. Hagger & M. Wilkin (Eds.), *Mentoring: Perspectives on school-based teacher education* (pp. 69–85). London, United Kingdom: Kogan Page.

McIntyre, D. J., Byrd, D. M., & Foxx, S. M. (1996). Field and laboratory experiences. In J. Sikula, T. Buttery, & E. Guyton (Eds.), *Handbook of research on teacher education* (2nd ed., pp. 171–193). New York, NY: Simon & Schuster Macmillan.

Peterson, B. E. (2005). Student teaching in Japan: The lesson. *Journal of Mathematics Teacher Education, 8*, 61–74.

Peterson, B. E., & Leatham, K. R. (2009). Learning to use students' mathematical thinking to orchestrate a class discussion. In L. Knott (Ed.), *The role of mathematics discourse in producing leaders of discourse* (pp. 99–128). Charlotte, NC: Information Age Publishing.

Rodgers, A., & Keil, V. L. (2007). Restructuring a traditional student teacher supervision model: Fostering enhanced professional development and mentoring within a professional development school context. *Teaching and Teacher Education, 23*, 63–80.

Stein, M. K., Engle, R. A., Smith, M. S., & Hughes, E. K. (2008). Orchestrating productive mathematical discussions: Five practices for helping teachers move beyond show and tell. *Mathematical Thinking and Learning, 10*, 313–340.

Wang, J., & Odell, S. J. (2007). An alternative conception of mentor-novice relationships: Learning to teach in reform-minded ways as a context. *Teaching and Teacher Education, 23*, 473–489.

Wilson, S. M., Floden, R. E., & Ferrini-Mundy, J. (2002). Teacher preparation research: An insider's view from the outside. *Journal of Teacher Education, 53*, 190–204.

Zeichner, K. M. (1981). Reflective teaching and field-based experience in teacher education. *Interchange, 12*(4), 1–22.

Zeichner, K. M. (2002). Beyond traditional structures of student teaching. *Teacher Education Quarterly, 29*(2), 59–64.

Keith Leatham is an assistant professor in the Department of Mathematics Education at Brigham Young University. His research focuses on the experiences that influence preservice secondary mathematics teachers' beliefs and knowledge about teaching mathematics. He can be reached at kleatham@mathed.byu.edu

Blake Peterson is professor in the Department of Mathematics Education at Brigham Young University. His research interests include Japanese mathematics teacher education and secondary mathematics student teaching. He can be reached at peterson@mathed.byu.edu

CPSIA information can be obtained
at www.ICGtesting.com
Printed in the USA
JSHW052044190323
39007JS00007B/11